Governing Affect

Governing Affect

Neoliberalism and Disaster Reconstruction

Roberto E. Barrios

University of Nebraska Press

Lincoln and London

Chapter 3 has been revised from "'Here, I'm Not at Ease':
Anthropological Perspectives on Community Resilience,"
Disasters 38, no. 2 (2014): 329–50. Chapter 5 has been
revised from "Malditos: Street Gang Subversions of
National Body Politics in Central America," *Identities: Global
Studies in Culture and Power* 16, no. 2 (2009): 179–201.
Chapter 6 has been revised from "'If You Did Not Grow Up
Here, You Cannot Appreciate Living Here': Neoliberalism,
Space-time, and Affect in Post-Katrina Recovery Planning,"
Human Organization 70, no. 2 (2011): 118–27.

Library of Congress Cataloging-in-Publication Data
Names: Barrios, Roberto E., author.
Title: Governing affect: neoliberalism and disaster
reconstruction / Roberto E. Barrios.
Description: Lincoln: University of Nebraska Press, 2017. |
Series: Anthropology of contemporary North America |
Includes bibliographical references and index.
Identifiers: LCCN 2016044827
ISBN 9780803262966 (hardback: alk. paper)
ISBN 9781496201904 (paper: alk. paper)
ISBN 9781496200143 (epub)
ISBN 9781496200150 (mobi)
ISBN 9781496200167 (pdf)
Subjects: LCSH: Disaster relief—Social aspects—Case studies. |
Natural disasters—Social aspects—Case studies. | Hurricane
Mitch, 1998—Social aspects—Honduras. | Disaster relief—
Social aspects—Honduras. | Hurricane Katrina, 2005—Social
aspects—Louisiana—New Orleans. | Disaster relief—Social
aspects—Louisiana—New Orleans. | Landslides—Social
aspects—Mexico—Grijalva River Valley. | Disaster relief—
Social aspects—Mexico—Grijalva River Valley. | Floods—
Social aspects—Illinois—Olive Branch. | Disaster relief—Social
aspects—Illinois—Olive Branch. | BISAC: SOCIAL SCIENCE /
Anthropology / Cultural. | NATURE / Natural Disasters.
Classification: LCC HV553 .B365 2017 | DDC 363.34/8—dc23
LC record available at https://lccn.loc.gov/2016044827

Set in Monotype ITC Charter by Rachel Gould.

For Teresa and Everardo

Contents

Illustrations

Acknowledgments

This book and my career as an anthropologist would not have been possible were it not for a number of people who supported and educated me over the course of four decades of life. First among these people are my parents, María Teresa Castellanos Pasarelli and Domingo Everardo Barrios Díaz, who made tremendous sacrifices to provide me with the resources and experiences that shaped me as a person and as a professional social scientist. Academics, of course, also have scholarly families, and I am forever indebted to the educators and researchers who encouraged me to pursue anthropology as a field of study. At the University of New Orleans, Malcolm Webb, Ethelyn Orso, Richard Shenkel, and Richard Beavers all played key roles in my mentorship and encouraged me to pursue a graduate degree in anthropology. At the University of Florida, Anthony Oliver-Smith, James P. Stansbury, Allan F. Burns, Lynette Norr, and Stacey Langwick were instrumental in guiding me to develop an interest in applied anthropology, the anthropology of disasters, and the anthropology of development, as well as providing me with the means necessary to complete my graduate training and dissertation research.

My graduate development would not have been the same were it not for a number of fellow students who volunteered their friendships and developing professional expertise, among whom I count Lauren Fordyce, Debra Rodman, Jennifer Hale-Gallardo, Sarah Graddy, Antonio de la Peña, and Robert Freeman. The ideas and research interests I developed in graduate school also came to fruition over the course of my postdoctoral employment as an assistant professor at Southern Illinois University–Carbondale (siuc), where a number of colleagues fostered the development of this book through casual conversations and helpful critiques. I am particularly indebted to Janet Fuller, Anthony K. Webster, Jonathan D. Hill, David Sutton, John McCall, René Francisco Poitevin, and Robert Swenson for making me a part of their intellectual communities and supporting my professional advancement.

Beyond the hallways of my home academic institution, I am also grateful to a number of fellow anthropologists who provided keen commentaries on my work that greatly enhanced the ways I think about post-catastrophe reconstruction. I am particularly grateful in this regard to Susanna Hoffman, Michele Gamburd, Mark Schuller, Faye Harrison, Katherine Browne, Rachel Breunlin, Helen Regis, Antoinette Jackson, David Beriss, Jeffrey Ehrenreich, Jesus Manuel Macías, Tang Yun, Zhang Yuan, AJ Faas, Gregory Button, and the two anonymous reviewers whose insights and recommendations greatly improved this manuscript. I would also like to thank James Bielo, Carrie Lane, Alicia Christensen, Sara Springsteen, and Vicki Chamlee at the University of Nebraska Press, who believed in and supported this book project.

The research upon which this book is based became possible through the kindness and hard work of a number of people who opened their homes, communities, and research sites to me. In Honduras, I was assisted by Marco Tulio Medina, Rosa Palencia, Carlos Giacoletti, Maria Fiallos, Myrna Portillo, and hundreds of disaster-displaced families who, despite the hardships brought about by forceful displacement and resettlement, always found time to share their life stories and thoughts with me. In New Orleans, Cheryl Austin, Jeanell Holmes, and Isabel Barrios Sherwood were instrumental in helping me understand life in the city before and after Hurricane Katrina. Qiaoyun Zhang, a graduate student from the SIUC Department of Anthropology, also contributed to my research efforts in southeastern Louisiana. In southern Illinois, Heather McIlvane-Newsad and David Casagrande were incredibly kind in inviting me to collaborate on their work on community resettlement and sharing their wealth of ethnographic data. I was also fortunate to be assisted by two undergraduate students, Katherine Ray and Christopher Higgerson. In Mexico, I remain indebted to Asunción Avendaño García, Roberto Melville, Virginia García Acosta, and the Centro de Investigaciones y Estudios Superiores en Antropología Social (Center for Research and Advanced Studies in Social Anthropology) for their hospitality and guidance. My ethnographic fieldwork was also facilitated by the Fulbright Institute of International Education program (which funded my dissertation research), the National Science Foundation (which assisted me in Honduras via a senior grant awarded to James P. Stansbury and Anthony

Oliver-Smith), and the siuc Faculty Seed Grant (which supported my work in New Orleans). Finally, I am eternally grateful to Kelly McGuire, who has been a friend and partner through this two-decade endeavor.

The research projects that compose this book would also not have been possible were it not for hundreds of people—disaster survivors, colleagues, government officials, nongovernmental organization project managers, planners, and architects—who kindly shared their time and thoughts with me. While some of my analyses of these interlocutors' ideas and practices are critical at times, I remain grateful for their trust, and my critiques are made in the spirit of enhancing recovery programs. To protect the privacy, reputation, and job security of my interlocutors, I have changed their names to pseudonyms, except in a selected number of instances when they explicitly requested that I use their real names.

Author's Note

Governing Affect is the result of four ethnographic research projects I conducted from 1999 to 2015. The research sites where I collected evidence include the greater Choluteca urban area in southern Honduras, various neighborhoods in the city of New Orleans, the town of Olive Branch in southern Illinois, and the resettlement community of New San Juan de Grijalva in the state of Chiapas, Mexico. Research in these four sites involved a range of methods including ethnographic interviews (casual and formal conversations with people about topics of ethnographic interest), participant observation (doing things alongside people) in community and institutional activities, household surveys, and the collection of anthropometric measurements (measurements of body size). With this evidence I make a number of claims about the challenges and unresolved contradictions of disaster mitigation policy and practice. To substantiate these claims, I use excerpts from ethnographic interviews and vignettes from participant observation activities throughout this manuscript. Anthropologists, I always tell my students, are very much like lawyers. We argue cases, and we must present evidence to the jury of our readers and fellow colleagues. The evidence must bear a logical connection to the claims we are trying to make, and it must be believable to the jury.

Anthropologists often differentiate between what they call unstructured, semi-structured, and structured ethnographic interviews. *Unstructured ethnographic interviews* are conversations anthropologists have with interlocutors that occur serendipitously as a result of the researcher becoming immersed in a particular community or institution and simply being in the right place at the right time. Unstructured ethnographic interviews are one of the most powerful and yet delicate forms of data gathering. The anthropologist must carefully and ethically balance the requirements of informed consent (reminding interlocutors that even though they may come to share bonds of friendship and even "fictive" kinship with the anthropologist, the anthropologist is still a researcher) and the trust and intimacy that their interlocutors demonstrate when, without solicitation, they pull the ethnographer aside and share intimate information about their lives and communities. While we learn the most in these moments, we are also given information that is sensitive and potentially harmful to the communities we study. The American Anthropological Association's ethical code stipulates that anthropologists must do no harm to the people and communities they study, and how we handle the information we are given during these seemingly casual exchanges can make the difference between doing something that is helpful and constructive and doing something that is incredibly harmful.

Because of the sensitive nature of information gathered through unstructured ethnographic interviews, ethnographers often wait until a conversation is over to write their journal entries, which anthropologists call *fieldnotes*. The anthropologist must attempt to re-create any exchange as faithfully as memory allows, and this task is not easy. Although unstructured ethnographic interviews may not seem a rigorous method to the uninitiated, their execution requires finesse and ethical awareness, and their documentation requires rigor and discipline. "You're just hanging out and talking to people, right?" someone might ask. Well, it involves more than that. Writing fieldnote entries after the fact is a time-consuming and intellectually draining task. Memory fades quickly, and details, phrases, and stories lose their resolution by the minute. The ethnographer must have the rigorous habit of writing things down as thoroughly as possible and as soon as circumstances allow.

Semi-structured ethnographic interviews differ from unstructured ethnographic interviews in that they are not unsolicited. In these instances, the ethnographer sets out to purposely have a conversation with an interlocutor about a topic of the former's interest. In some instances, the anthropologist may take brief notes during the conversation and later flesh them out as more detailed fieldnotes, with the end result being a journal entry that resembles those created for unstructured interviews as well.

By contrast, *structured ethnographic interviews* are more formal affairs. The anthropologist often brings lists of topics of conversation to ensure she or he covers them with their interlocutors. The researchers may also take more detailed notes than they would during a semi-structured interview and, only with the explicit and documented authorization of the interlocutor, may even create an audio recording of the exchange if circumstances and cultural norms allow. While these latter ethnographic interviews may seem superior due to their structure and documenting techniques, they may not be as rich in information because the interlocutor is made overly aware of the research process. These interviews may therefore elicit "official" versions of events and opinions, whereas unstructured ethnographic interviews are critical moments that occur in practice during which the interlocutor demonstrates a cultural phenomenon or shares information whose relevance the anthropologist may not have known about and was therefore incapable of asking questions about it. As a rule of thumb, what does not look rigorous in ethnographic research is quite the opposite, and what seems most rigorous may be that which reveals the least.

In *Governing Affect*, I differentiate evidence I gathered through unstructured, semi-structured, and structured ethnographic interviews with formatting that separates it from the rest of the text and by adding a citation that indicates the year I conducted the interview and the manner in which I handled the information. For example, information collected from a semi-structured ethnographic interview in 2011 features a parenthetical citation at the end of the given section that reads "(semi-structured interview 2011)." A structured ethnographic interview that I conducted in 2008 and that I audio recorded and later transcribed, in contrast, reads "(structured interview transcription 2008)." If I did not audio record and transcribe this latter interview, I then cite it as "(structured interview 2008)."

Complementing interviews, *participant observation* is the other mainstay of ethnographic research. One of the potential pitfalls of ethnographic interviews is that they can often elicit "official" representations of events and people that may not match what people actually do in practice. Anthropologists therefore supplement their interview materials by developing profound rapport with their cultural interlocutors and doing things alongside them. In doing things with people, anthropologists can capture what their interlocutors may not be able to speak about but can demonstrate in action. As with their unstructured and structured ethnographic interviews, anthropologists rely on short-term memory and the writing of fieldnote entries to process the information they gain from participant observation activities. In this book, I include multiple excerpts of comments my interlocutors made during participant observation activities that I documented in the form of fieldnotes. When I use fieldnote excerpts as evidence, I also separate this evidence from the remainder of the text and close the section with a citation that notes the year in which the activity and documentation took place. For example, evidence detailing the ways expert planners spoke about the recovery process in New Orleans during a 2006 planning meeting where I conducted participant observation is cited as "(fieldnotes 2006)."

Introduction

AFFECT AND EMOTIONS IN

DISASTER RECONSTRUCTION

On a hot and humid summer day in 2009, Ward "Mack" McClendon agreed to sit with me outside of a large green warehouse located in the Lower Ninth Ward—a part of New Orleans devastated by Hurricane Katrina's floods—to talk about his assessment of the area's recovery. Before the storm, Mack dedicated himself to restoring old cars and driving a tow truck, bringing in a comfortable income. The disaster and the way local and federal government agencies handled the area's reconstruction, however, resulted in the partial disappearance of what he had come to take for granted in the preceding years: his friends, neighbors, acquaintances, and relatives, as well as their particular ways of speaking, behaving, socializing, sharing food, and everyday ways of being that generated a sense of comfort and wellness for Mack.

In 2009 only 15 percent of the area's pre-Katrina households actively received mail, a proxy measure demographers used to estimate the rate of population return after the hurricane. By 2015 this number had increased to 37 percent, while the citywide figure had risen to 90 percent (Allen 2015; Plyer and Mack 2015). The absence of familiar faces and embodied ways of being struck Mack in what social scientists would label *an affective way*. Mack felt this absence; the feeling he experienced was an uneasy sense of loss that drove him to do things he never considered doing before the catastrophe. Mack reflected:

> Believe it or not, before Katrina, I was a very private person, okay, but my community is hurting so bad, I can never be the same person I was before. After embracing the problems that we have, you got to change, and it's a good way. It's not a bad way; it makes you start caring about people. (structured interview transcription 2009)

Over the years I knew Mack, I also came to recognize his concern for how those people who were not born and raised in the Lower Ninth Ward but had come to help rebuild after the catastrophe cared about the neighborhood. In some instances, academics, environmental activists, and non-profit program managers seemed more worried about such things as the salinity levels of nearby wetlands and the energy efficiency of homes with low carbon footprints than about the New Orleanians who lived in the area before the flood. Mack felt these residents were irreplaceable, even as new arrivals from other parts of the city and the United States created the impression that the neighborhood was slowly "coming back."

Mack's concern about the Ninth Ward's reconstruction moved him to do something he would have considered illogical before Katrina: he gave up his towing business and dived head deep into community organizing. With his own finances, he purchased a warehouse located in the flooded neighborhood and remodeled it as a community center where out-of-town reconstruction volunteers and residents who needed assistance with home repair could connect. This decision was financially difficult for Mack, and he faced great challenges over the next five years, including the death of his daughter during childbirth and the bank's foreclosure on his house. But he never questioned his decision in my presence. Mack's emotion-laden response to the absence of people and practices he found culturally familiar had affected him in such a way that community organizing was something he *had* to do.

One thing that struck me about our conversation was that it was not the first time in my decade of ethnographic research that a person who had lived through a disaster used a language of affect and emotions when assessing his or her community's recovery process. Nine years before my conversation with Mack, the people of Choluteca, Honduras, who were displaced by Hurricane Mitch's floods also alluded to their bodily sensed notion of comfort (which was triggered by the spatial proximity of trusted friends, relatives, and familiar architectural structures) as the criterion by which they reflected on the merits of governmental and nongovernmental organization (NGO) reconstruction programs. New Orleans would also not be the last place where I would hear my ethnographic interlocutors make such statements. In the years and research projects to follow in the midwestern United States and Chiapas, Mexico, people I spoke with would similarly reference their sensed and emotion-laden experience of neighbors, relatives, and spaces as the mechanism by which they evaluated disaster risk and recovery.

On the day of my conversation with Mack, however, I could not foresee that I would one day be sitting down to write this particular book. Disasters, after all, are often represented in popular media as states of emergency in which pragmatic decisions concerning life and death must be made on the fly, while emotions are viewed as sensory experiences whose consideration requires a slowing down of practice and as a luxury that can only be afforded by those not facing an imminent geophysical threat or the widespread disruption of a catastrophe. I would also have had difficulty understanding the relevance of research on emotions and affect for disaster survivors and the myriad professions involved in disaster reconstruction. Yes, the impact of catastrophes on built and "natural" environments is one that usually makes a significant emotion-evoking impression on television audiences, and news media outlets are all too eager to exploit its sensationalist potential. Emotions, one could say, are "all over disasters." Nevertheless, as we shall see throughout *Governing Affect*, disaster mitigation experts often dismiss the more mundane feelings (e.g., people's attachments to small rural towns that have seen better days, to socioeconomically disadvantaged neighborhoods, and to social relations with friends and relatives) of those who directly experience devastation as obstacles to the application of rational best practices in disaster prevention and recovery.

It is noteworthy that a number of anthropologists have documented the ways emotions manifest in disaster contexts as public reactions to sensationalist news or propagandistic state coverage (Makley 2014), as grounds for identity formation on the basis of shared suffering (Oliver-Smith 1986), and as movers of collective action in the form of volunteerism and personal donations (Adams 2013). I also recognize that, recently, Katherine E. Browne (2015) has begun to take a closer look at the relationships between comfort and kin relations in post-disaster contexts and at the importance of this web of practice, sociality, and feeling in the recovery of communities. This book, however, is about the ways people who live through disasters invoke emotions as a means

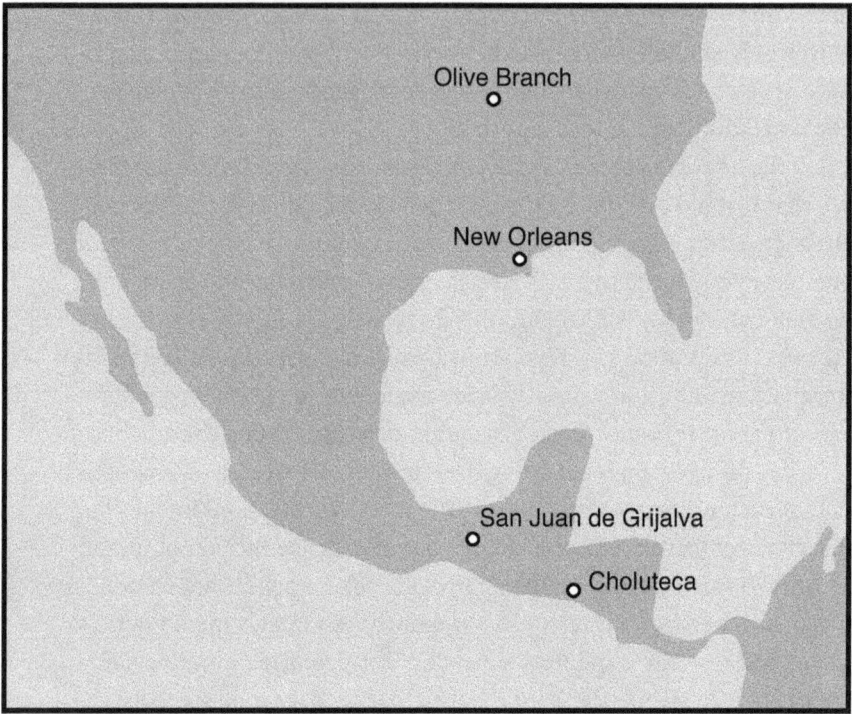

FIG. I. Research sites. Courtesy of author.

of assessing the relevance of governmentally sanctioned recovery plans, judging the effectiveness of disaster recovery programs, and reflecting on the risk of living in areas that have been deemed prone to disaster hazards. Affect and emotions, I claim, are by no means irrelevant to the study of disasters and the distribution of reconstruction aid. The cases I present from southern Honduras following Hurricane Mitch, New Orleans after Hurricane Katrina, Olive Branch in southern Illinois after the Mississippi River flood of 2011, and San Juan de Grijalva in the state of Chiapas, Mexico, after the 2007 Grijalva River landslide (fig. 1) demonstrate that feelings are central to people's experience of catastrophes and recovery. They must therefore be carefully apprehended, considered, and addressed by those interested in enhancing post-disaster assistance and risk reduction.

While taking affect and emotions into account in disaster prevention and recovery may seem initially like a simple task, the topic is more complex. First of all, a critical reader may ask why I have chosen to focus simultaneously on both affect and emotions. What do I mean by the two terms, and why do I mention both as if they were separate and distinct phenomena? As I explore in greater detail in chapter 1, these questions reflect a long-standing dialogue in the humanities and social sciences. Immanuel Kant (1996 [1797]), for example, saw affects (in the plural) as differing from passions, both of which he considered subcategories of emotion. In Kant's categorization, affects (e.g., anger, lust), on the one hand, precede reflection; they are quasi-involuntary reactions to social situations and experience. Passions (e.g., hatred), on the other hand, are "a sensible desire that has become a lasting inclination" (1996, 208) and are therefore subject to reflection.

In other instances, anthropologists such as Frances Hsu (1977) have used affect and emotion interchangeably, suggesting that the two terms are synonymous. More recently Brian Massumi (2000) has once again distinguished affect from emotion. He uses the former term to indicate bodily reactions to external stimuli that do not enter a person's consciousness and the latter to refer to a sensory experience that a person becomes aware of and interprets in a culturally particular way, or what he calls a "socio-linguistic fixing of the quality of an experience which is from that point onward defined as personal" (Massumi 2000, 88).

Suffice it to say for now (see chapter 1 for a more thorough discussion), in this book, I use the term *affect* to refer to a sensory experience that is felt by a body in relation to another, human or otherwise (Seigworth and Gregg 2010; Spinoza 1994), and I understand the body that feels as a product of human practice and human-environment interactions, or as an embodied way of being. I use *emotion* to refer to affective experience as it is narrativized by people, structured in a culturally particular way, and put to a political or social use—for example, what or whom to love, hate, or fear and how (Lutz 1986, 1988; Abu-Lughod and Lutz 1990). As I show through various ethnographic examples, affect is a sensory experience that disaster survivors often attempt to apprehend linguistically, therefore crossing the threshold that separates it from emotion in Massumi's definition. I argue that the dramatic impact of disasters on the built, natural, and social environments, on whose presence or remembrance affective experience is contingent, presents a unique circumstance that drives disaster survivors to reflect on what is often sensed but is not necessarily brought into discourse. This linkage, as Mack McClendon tells us, can become a driving force of social change, leading people to do things and become persons they otherwise would not have done or been.

Another complication of writing about affect and emotion is that people often naturalize the bodily experiences of disgust, fear, comfort, or desire as if they were the manifestation of biologically determined (isn't it natural to be frightened by x?), rational (isn't it logical to be disgusted by y?), or universal (doesn't everyone desire z?) ways of reacting to others and things. The anthropological literature, however, is replete with examples that tell us something quite different. Emotions and sensory perception can vary tremendously from one cultural context to the next, and what a person experiences as grotesque or comforting in one setting may not be considered so by someone in another. In fact, sensory and emotional experiences were one of the primary ways through which people experienced cultural difference and enacted ethnocentrism in colonial situations (Povinelli 2002, 2006), and they continue to be a primary way of making and maintaining race and class distinctions (Stewart 2007).

The anthropological literature, then, requires us to ask the question, if emotions and affect are differently experienced, then what evokes an emotion or bodily reaction for whom and why? The cases I present demonstrate that people come to experience sensory and emotional states in unique ways as a result of their life experiences in specific contexts, ones that are material, cultural, political, and environmental at once. Affect has both historicity and ecology, meaning that bodies are not given in nature with a predetermined or hardwired way of sensing the world and relationships around them; instead, they emerge in relation to socially structured and meaning-laden relationships with people and things in what I call an ecology of affect. A concern with affect, then, bridges the gap between meaning and materiality and collapses a number of binary representational conventions that limit anthropological analyses of people and cultural practice: nature versus culture, subject(ivity) versus object(ivity), and static "traditional" past versus changing present. This brings about yet another complication that preoccupies this book: if bodies and their sensing and emoting capacities are not given in "nature" but emerge in affective ecologies, then they may always be in a process of emergence, and how can an ethnographer document (much less make policy recommendations about) something that is forever in a process of becoming?

Through the case studies featured in this book, I hope to show that affect is simultaneously emergent and mnemonic—a term I use to denote how the body's sensing capacities can conjure memories through the detection of familiar people, objects, smells, and tastes (Jackson 2011; Navaro-Yashin 2012; Proust 2006; Sutton 2000). By the same logic, affect is neither static nor unchanging: what is first unfamiliar and unpalatable may become recognizable and pleasant (or vice versa) depending on the social relations within which it is experienced. Nevertheless, bodies are not computer hard drives whose memories can be easily erased or whose "programs" can be simply and predictably rewritten.

The simultaneously emergent and mnemonic qualities of affect can be particularly challenging in the case of disaster recovery. Disasters and reconstruction programs often radically transform the social relations and built environments that evoke familiar and comforting affective reactions (Oliver-Smith 1986, 2002; Ullberg 2013). If disaster survivors mobilize a language of affect when assessing the relevance of disaster reconstruction projects or gauging progress toward recovery, how can aid program managers and disaster recovery planners re-create a world that may no longer be feasible? At the same time, what are policymakers and program managers to do when disaster survivors cannot experience ease or security in the socio-spatial arrangements of reconstructed or resettled communities? If affect is emergent, why can't some disaster survivors simply get used to a new state affairs?

It is worth noting that there is a growing and important body of literature on memory, identity, and disasters (Doss 2010; Gray and Oliver 2004; Simpson 2013; Ullberg 2013). This book distinguishes itself from these other works by focusing on the mobilization and invocation of affect among disaster survivors. This practice has mnemonic dimensions, but as I further explore in chapter 1, it also merits analysis from the vantage point of the anthropology of the body and practice theory (Bordo 1993; Bourdieu 1977; Farquhar 2002; Lock and Farquhar 2007; Stoler 1995).

Governing Affect is also about the ways urban planners, NGO program managers, and governmental officials involved in disaster mitigation implicitly figure sensed experience and emotions in policy and institutional practice, which are often inflected with neoliberal and modernist assumptions about the natures of people and well-being. In the presented case studies, I show how disaster recovery experts and political elites often render the emotions and attachments of subaltern (a term I use to describe people who find themselves in a condition of sociopolitical subordination) disaster-affected populations as obstacles to fiscal cost-benefit analysis or techno-scientific disaster management, while at the same time the experts and elites promote the desire for built environments and human-material relations, which they credit with the capacity to reproduce capital or shape normative human behavior.

Like affect and emotions, the terms *neoliberalism* and *modernity* have long histories of examination, discussion, and debate in the social sciences. For the sake brevity, allow me to briefly clarify what I mean by each, understanding that such cursory treatment leaves out the overwhelming majority of volumes on these topics. At the same time, rest assured that each of the succeeding chapters further engages the existing scholarship on these two concepts.

My use of "neoliberalism" is informed by Michel Foucault's (2004) and Elizabeth Povinelli's (2010) recognition of a cultural trend where policymakers, political leaders, and the public at large propose the subjection of all facets of human life to capitalist cost-benefit analysis as a mechanism for creating social well-being. Related to this particular kind of political imagination is the idea that market liberalization (e.g., the deregulation of labor, environmental policy, and financial markets) will lead to optimal social ends (di Leonardo 2008). Social scientists who trace the history of neoliberalism often see its emergence as a response to the global capitalist crisis of the 1970s, although neoliberalism, just as all types of global flows, is a changing entity that people interpret and reconfigure in varying ways from one locality to the next. At the same time, research on disaster recovery has demonstrated that rather than operating as a complete retraction of government, disaster neoliberalism works more toward the rearrangement of relationships between government and the private sector in a particular type of corporatism where the former channels public funds and resources to the latter in exchange for the provision of services (e.g., disaster aid and case management) (Adams 2013). Neoliberalism, then, is not singular and homogenous but plural and mutated, and disaster contexts, with their global distribution, are particularly interesting sites to explore its proliferation and diversification.

By "modernity," I mean to call attention to ethnocentric ways of thinking and governing where cultural difference is implicitly rendered in hierarchal temporal terms. Specifically I mean those cases where policymakers and sociopolitical elites figure the cultural practices of people that are deemed "other" in their national imaginaries (e.g., indigenous, racialized or ethnicized, and subaltern populations) as vestiges of prior developmental stages of a linear evolutionary history (Fabian 1983; Povinelli 1995). Of particular relevance here is the way anthropologists interested in modernism have documented how some technocrats envision specific modifications of the built environment (e.g., spatial homogenization and regimentation, postmodern aesthetic design) as mechanisms for transforming "poor" or "traditional" peoples into the kinds of subjects who populate neoliberal and modernist imaginaries (Caldeira and Holston 2005; De Cunzo 2001; Holston 1989; Rabinow 1995).

What specifically concerns *Governing Affect* is how politically and hegemonically influential actors often seize post-disaster contexts as opportune moments for bringing about the dramatic transformation of urban and community spaces under the auspices of "rebuilding better" and how accompanying definitions of "the better" seem to repeatedly entangle neoliberal and modernist assumptions about the natures of people and the common good. It is noteworthy that a significant body of literature recognizes how disasters have increasingly become an advantageous moment to carry out or expedite dramatic transformations of economies, cities, and nations along neoliberal principles (Adams 2013; Button 2010; Button and Oliver-Smith 2008; Gunewardena and Schuller 2008; Klein 2007; Rozario 2007). What distinguishes this book is its focus on the ways neoliberal and modernist tenets are entangled with existing social orders that are a long time in the making (i.e., postcoloniality), on the unique and contingent ways they are interpreted and reconfigured across disaster-affected localities, and on the affective frictions they elicit.

By looking at two intersecting points of interest—the impacted populations' affective experience of reconstruction following catastrophes and the figuration of affect in modernist and neoliberal recovery policy and practice—this book explores a number of complications, tensions, and mediations that characterize the ways sensory perception and emotions manifest in disaster contexts. As Mack's case demonstrates, affect is both a primary mover of social action (feelings of loss, desire, love, or fear move people toward particular ends) and a fundamental dimension of human experience; inequity, vulnerability, and recovery are conditions that are, first and foremost, *felt* (Fassin 2013; Seigworth and Gregg 2010). *Governing Affect* shows how disaster recovery practices on the part of assisting governmental and nongovernmental agencies that ignore the felt experience of disaster survivors run the risk both of failure in practical terms and of being perceived by affected populations as culturally insensitive and disruptive, if not ethnocidal. An affect-centered approach to disaster recovery, I argue, is key to adapting governmental and NGO reconstruction policies to the embodied cultural particularities of the people who live in catastrophe-affected sites.

An Ethnographer's Journey, a Book's Roadmap

The process through which I came to recognize the importance of affect and emotions in disasters was not a straightforward one. This trajectory was one part biographical, one part corpus of anthropological literature, and one part collection of ethnographic experiences as a disaster researcher. My path illustrates how the production of anthropological knowledge is influenced by the ethnographer's life history, which shapes the researcher as a particular kind of person with unique interests, passions, politics, interpretations of the anthropological canon, and ways of seeing and processing the world. At the same time, this kind of knowledge making is also influenced by the ways the ethnographic method has a feedback effect on anthropologists, transforming how they engage and understand their field experiences. Ethnography involves a co-constitutive relationship between the producer of anthropological knowledge and the people and places that the researcher studies.

My journey into the field of disaster research grew out of an interest I developed in graduate school about forced population displacement and resettlement, a topic I chose for rather personal reasons. Born in Guatemala, Central America, I experienced my own form of involuntary migration at the age of thirteen when my family moved to New Orleans, Louisiana, in 1987 to distance itself from escalating state violence and civil unrest. The process of transnational migration was not an easy one for me. Prior to our permanent move, my exposure to life in the United States was limited to holiday visits with relatives who migrated there in the 1970s and to touring amusement parks, tourist attractions, and shopping malls. Through these visits, I came to imagine the United States as a space of consumption and gratification, a haven from the sociopolitical troubles of Central America.

My experience of an idealized North America was very different from the life I came to know as a transnational migrant in Metairie, a New Orleans suburb. In this new context, I came to know a United States with decrepit apartment complexes, anonymous neighbors who remained strangers despite spatial proximity, and stark socioeconomic inequities between low-level service sector workers and the financially accommodated suburban middle class. Perhaps most important, I also came to know a new world of racialized difference where classmates, teachers, and neighbors perceived me as someone unlike them, as a "Latino" who was regarded with curiosity, condescension, and sometimes suspicion but seldom as an equal.

The experience of being "othered" as a Latino was difficult, but it was also educational. In my prior life in Guatemala, I had the privilege of being ethnicized, racialized, and gendered as a Ladino—a term used to denote people who claim "mixed" European and indigenous ancestry although they engage in behaviors that many Central Americans consider indexical of an affinity for European and North American culture and modernity (Nelson 1999a, 1999b). As a male Ladino, I could make claims to membership in a normative social category that many of my compatriots believed epitomized hegemonic nationalist ideals of citizenship, race, and class. Although people who identify as Ladinos in Guatemala are by no means a socioeconomically homogeneous or egalitarian group (never underestimate the capacity of dominant groups to discriminate among themselves), the privileges I enjoyed over the course of my childhood and early adolescence vanished in Metairie. The experience of being othered as a Latino brought a body politic—a term I use to denote the ways people attempt to structure societies along socially produced racialized, ethnicized, and gendered distinctions—into relief for me that had been previously invisible.

The recognition of the historical and cultural contingency of social orders in Guatemala and New Orleans required me to reflect on my previous formative years in Central America. My migration experience led me to reconsider the way Guatemalan society was structured: my fellow Central Americans associated Ladinos with modernity and progress and saw them as worthy of benefitting from the free or underpaid labor of indigenous populations, while they also represented indigenous communities as obstacles to national development (Nelson 1999a; Way 2012). The realization that such a structuring of society was not a natural or logical order but the result of a history of race and ethnic relations dating to the sixteenth-century colonization of Central America and that the inequities engendered by such relations caused the turmoil that forced my family to migrate to the United States proved to be a strong motivation for pursuing anthropology as a course of study at the University of New Orleans. To put it another way, I was driven by simple questions that required complex answers: Why do people make certain worlds and not others? Why do categorizations and hierarchies that are primarily historical and social in origin come to be seen as a natural and necessary way of structuring societies? I would return to these questions time and time again over the course of my disaster research in Honduras, Louisiana, Illinois, and Mexico.

While my migration experience translated into an intellectual practice of questioning social orders, its accompanying displacement became a sensory and emotional one. I didn't just *think* about my newfound status as an ethnicized and racialized other in Metairie; I *felt* this experience on a daily basis. Foods tasted differently and had unfamiliar textures, and the tastes and feelings I had once found comforting and familiar in Guatemala were difficult, if not impossible, to find. Environments, both built and natural, had different smells and appearances and thus had different emotional impacts on my body. I longed for the urban landscape of Guatemala City, where I had grown up (and I am, today, always taken aback when fellow North Americans who have visited this metropolis remark on the ugliness of its urban landscape), as well as the mountainous panoramas of my previous home that were nowhere to be found in the predominantly flat lower Mississippi Delta. I also experienced anxieties about having to interact with other adolescents whose official and unofficial rituals of socialization and performance of "coolness" were unknown to me.

The suburban space that I had once associated with the short-lived thrill of commodity consumption and the comfort of seeing relatives during holiday visits turned into a space of discomfort and anxiety, where the thought of having to navigate race and ethnic relations and institutionalized spaces such as junior high schools made parts of my body tighten with apprehension. As it turned out, my research on post-disaster reconstruction would lead me to investigate the very relationship between built environment, sociality, sensory perception, and emotions I had lived through, but could not put into words, as an adolescent.

Upon receiving my BA in anthropology, I enrolled in the graduate program at the University of Florida. Under the direction of Allan F. Burns, I decided to focus my master of arts project on Guatemalan Maya diaspora communities in South Florida that their members formed in the early 1980s after escaping genocide in Central America (Burns 1993). My ethnographic work with these communities documented the ways transnationally displaced Kanjobal Maya people created new spaces for themselves in North American society and generated a sense of cultural continuity from one generation to the next. My research looked at the ways Kanjobales organized patron saint festivals; circulated musical instruments, textiles, and foods in exchange networks that cut across national boundaries; and emphasized Maya language education, with the effect of producing a collection of identity-forming and disposition-cultivating life experiences. Kanjobales in South Florida were involved in making spaces that both evoked familiar sentiments in a dramatically new context and shaped the emotion-laden experience of Kanjobal language, material culture, and daily rituals for new generations.

In 1998 I expanded my work to include other Kanjobal refugee communities in Chiapas, Mexico, but it was an uncertain time to work in the region. During my preliminary research, Mexico's military established roadblocks that cut access to several key refugee communities during its counterinsurgency campaign against the Zapatista Army of National Liberation (EZLN). In Chiapas staff members of the United Nations High Commissioner for Refugees who had originally agreed to facilitate my work became less and less enthusiastic about taking responsibility for a graduate student's moving around and asking questions concerning forced displacement in a militarized zone. Back in Florida, my graduate school advisers also worried about my dissertation project's feasibility and strongly encouraged me to develop a new research focus.

In late October 1998 a category 5 hurricane made its way through the Caribbean, leaving devastation in its wake. Hurricane Mitch eventually reached the north coast of Honduras, unloading a deluge over the country. Centuries of deforestation, environmental degradation, inequitable land tenure relationships, stark social inequities, extractive colonial and postcolonial global economies, and unplanned urban growth—all came together with this storm to create one of the most catastrophic disasters of the twentieth century. In Honduras alone, national public health agencies reported 6,500 people dead, 8,000 missing, and 35,000 homes destroyed (AMDC 1999; PAHO 1998). At the University of Florida, under the joint direction of Anthony Oliver-Smith and James P. Stansbury, a group of faculty members and students put together a research project focused on documenting the reconstruction process and using anthropological theories and methods to enhance aid delivery.

The University of Florida group's focus on disaster-induced displacement resonated with my interest in refugee communities, and in the summer of 1999, I traveled to Honduras for the first time to complete a brief survey of child nutritional status in three hurricane-affected areas of the country. This pilot project, which was directed by James P. Stansbury and funded by the International Hurricane Research Center at Florida International University, consisted of collecting child-growth measures for three hundred children younger than five years of age from three hurricane-affected populations. The surveyed groups included people living in a large-scale temporary shelter of the capital city, Tegucigalpa; residents of hurricane-flooded neighborhoods of the western town of Catacamas; and settlers of the largest housing relocation site in southern Honduras, Limón de la Cerca.

The anthropometric measures I collected with the help of my research assistant, Rosa Palencia, suggested that children younger than five years of age in Limón de la Cerca were experiencing unprecedented levels of acute malnutrition (Barrios et al. 2000). In 2000 I returned to southern Honduras with funding from the National Science Foundation (acquired by James P. Stansbury and Anthony Oliver-Smith) and the Fulbright International Institute of Education to complete a more extensive nutritional survey and a thirteen-month ethnography of the reconstruction process in Limón and a second nearby reconstruction community called Marcelino Champagnat. The intention behind the nutritional study was to use child-growth measures and household dietary recalls as a means of assessing the long-term impacts of disasters and reconstruction on community health.

In addition to nutritional epidemiological methods, my ethnographic research involved participation in community activities, spending as much time as possible in the reconstruction site, befriending disaster survivors, and conducting ethnographic interviews with residents, NGO program managers, and local government officials. Through these latter activities I noticed and documented social conditions manifested in Limón and Marcelino that the nutritional instrument's narrow focus did not allow me to record but were nevertheless of great importance for understanding the reconstruction process.

Over the course of its (re)construction, Limón de la Cerca gained a reputation as a place of street gang activity, intra-community violence, and social fragmentation. The housing structures built by donor organizations were poorly designed, did not suit the needs of disaster survivors, and were not constructed to withstand the environmental stresses of the site. Living conditions were crowded; homes did not feature the separations between dining, sleeping, and visiting quarters that Cholutecans were accustomed to; and housing structures either collapsed or lost their roofs during heavy rainstorms, causing severe injuries and deaths. The structures were also located on minimal land parcels that limited practices that displaced southern Hondurans had engaged in before the storm: expanding their homes to meet family needs, planting house gardens, or raising pigs and chickens.

Over the course of my ethnographic interviews, I became interested in the ways disaster survivors assessed the efficacy of reconstruction practices on the part of the local government, the national and international NGOs, and their own grassroots leaders. As I document in chapter 2, the disaster-displaced Hondurans I spoke with made such assessments in terms of a local colloquialism, *hallarse*, which they used to convey "the sense of finding oneself at ease." *Hallarse*, these disaster survivors told me, was an affective experience, or something one felt. This feeling was evoked by the presence of trusted neighbors, the proximity of relatives, and the spatial conditions that sustained such a social and affective landscape—that is, ample land parcels that allowed future home expansions as families grew from one generation to the next and the distribution of disaster survivors that permitted trusted neighbors and relatives to live close to each other. The socio-material conditions that formed over the course of Limón de la Cerca's (re)construction, I was told on multiple occasions, did not allow disaster survivors to experience *hallarse*.

The grim assessment of life in Limón on the part of displaced Cholutecans did not match the reports and evaluations produced by the local government and the U.S. Agency for International Development (USAID), the latter having partly funded the site's construction. Local government reports heralded the reconstruction of Limón de la Cerca as exemplary of successful participatory development. Titles such as "Participation Towards Reconstruction" accompanied photographs of neatly aligned and diminutive houses in local government annual reports, and USAID program evaluators were happy to report that reconstruction program budgets were spent on time on their designated purposes. Reconstruction, as far as these agencies were concerned, was moving along splendidly.

Over the course of my fieldwork in Honduras, I learned the policies and practices of NGOs and local governments warranted interpretation as statements concerning the nature of social well-being; such statements had intricate and unique histories. These histories, in turn, connected distant places and imaginations across the globe. Designed by Nacional de Ingenieros, one of Honduras's premier engineering firms, the layout of Limón de la Cerca's housing plan implicitly articulated an

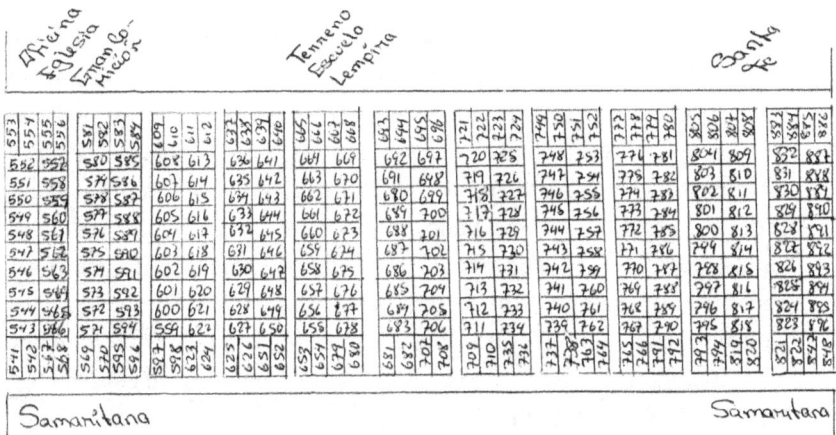

FIG. 2. A detail of Limón de la Cerca's master plan showing the site's regimentation and homogenization of space. Courtesy of the Honduran Institute of History and Anthropology.

idea about the relationship between rigidly regulated space and personhood—specifically, that uniformly arranged land parcels and homes can produce normatively behaving people (fig. 2). Anthropologists such as Paul Rabinow (1995) and James Holston (1989) trace the history of this notion to the birth of French modern urbanism in the nineteenth century.

Likewise, the evaluation techniques of USAID, which emphasized the role of budgets as a means of making knowledge about recovery programs, were not executed in a cultural or historical vacuum. Instead, they were indicative of the kind of management practice Frederic Jameson (1991) associates with recent stages of capitalism and that Michel Foucault (1991) considered techniques of modern governance—what he calls governmentality—that came about in western Europe between the eighteenth and nineteenth centuries. Finally and perhaps more important, my time in Honduras also exposed me to reconstruction sites such as Marcelino Champagnat, where NGO program managers and disaster survivors negotiated different ways of managing and implementing reconstruction projects and created social and material circumstances that evoked sentiments of recovery among the latter.

Processing my experiences in Honduras would prove a challenging but transformative task. My initial interest in nutritional epidemiology and anthropometry as a mechanism for documenting the long-term impacts of disasters seemed too narrow a way of relating the political and epistemological challenges of disaster recovery—that is, the techniques through which NGO staffs evaluate aid programs and the fundamental assumptions about people and well-being implicit in recovery plans and projects. The nutritional epidemiological information I recorded, for example, limited my statistical analysis to disaster survivor households and did not allow me to explore the broader relationships between the displaced Cholutecans, the NGOs, and the local government that I had witnessed, much less to investigate the implications of modernization, governmental rationality, and connections between local, national, and international politics in disaster recovery.

The case of post-Mitch reconstruction illustrated how, contrary to official representations of international NGOs and local government officials, aid practices that sought to modernize and develop disaster-affected communities with little regard for the socio-spatial arrangements that the survivors use to create a sense of place (the varied meanings, memories, and affective experiences that people associate with specific spaces) did not alleviate the long-term impacts of the disaster but rather exacerbated them. As Susanna Hoffman and Anthony Oliver-Smith (1999) have observed, disasters are processes engendered by long-standing human-environment relationships that enhance a geophysical phenomenon's materially destructive and socially disruptive capacities and unequally distribute a catastrophe's effects along the fault lines of a society's body politic (e.g., socially produced gender, race, ethnic, and class differences). What is more, the disaster as process does not end with the receding of floodwaters or the cessation of seismic movement, and it can extend indefinitely and descend into everyday life (Hastrup 2011), depending on the ways NGOs, governmental organizations, and disaster-affected communities handle the reconstruction process (Bankoff and Hilhorst 2004; Schuller 2012; Tierney and Oliver-Smith 2012).

Over the course of the succeeding years of conducting presentations, reviewing the existing literature, and listening to the research findings of fellow social scientists interested in post-disaster recovery, I have also discovered that what transpired in Limón de la Cerca was not an anomaly but the norm of reconstruction practices on the part of many governmental and nongovernmental agencies. Case studies from Turkey (Ganapati and Ganapati 2009), Mexico (Audefroy and Cabrera 2014; Briones 2010; Macías 2009), and other regions of Honduras (Snarr and Brown 1979) offer eerily similar scenarios in which community reconstruction projects disregard the spatialized and affect-laden kin and neighborly relationships among disaster survivors and restructure the space-times of their communities along rubrics of modern urbanism. While unique, the case of Limón de la Cerca also seemed more and more to be a suitable point of departure for examining broader issues concerning space, affect, sociality, and governance at stake in disaster recovery.

While chapter 2 focuses on the comparison of Limón de la Cerca and Marcelino Champagnat—two community resettlement sites that followed dramatically different paths over the course of their (re) construction—chapter 3 details the ways disaster survivors differentially experienced recovery projects along gender lines and how these projects themselves created new ways of experiencing the gendered self. Chapter 4, in turn, takes a close look at the ways state and news media attempted to cultivate affective reactions such as terror and disgust among the Honduran populace in relation to unruly subaltern adolescents involved in street gangs, which proliferated in the disaster's aftermath. Tracing such cultivations of affect, I propose, is a resourceful methodology for understanding both the ways power operates through sentiments in postcolonial contexts and the ways power is contested by the emotion-laden attachments that characterize the socialities and kin relations of socioeconomically disadvantaged Hondurans.

Not long following the completion of my work in Honduras, another disaster steered my journey as an anthropologist in an unexpected direction. In late August 2005 I received phone calls from my parents and sister informing me they were evacuating from the city of New Orleans before Hurricane Katrina made landfall in the U.S. Gulf Coast. Their evacuation was a tedious one. Bumper-to-bumper traffic along westbound I-10 turned a six-hour drive to the city of Houston into a twenty-two-hour ordeal. It marked the beginning of a six-month displacement that resulted after New Orleans's levee system, which was built and "maintained" by the U.S. Army Corps of Engineers and a number of independent contractors and local levee boards, failed in multiple locations and catastrophically flooded 80 percent of the city.

Following my dissertation research, I planned to steer my work away from disaster studies and focus on adolescent street gangs in Central America. In Katrina's aftermath, friends, mentors, and colleagues inquired whether I planned to do what seemed obvious to them and begin another post-disaster ethnographic project. In June 2006, after being hired as an assistant professor of anthropology at Southern Illinois University–Carbondale, I traveled to Louisiana and spent the first of many summer, spring, and winter breaks there with the intention of documenting its recovery.

My research in Honduras had taught me that reconstruction policies and practices on the part of governmental and nongovernmental organizations always carry implicit assumptions about the natures of people and communities, and these assumptions do not necessarily map neatly onto those relationships among people and between people and things that evoke sentiments of recovery and well-being among disaster survivors. I began my research, then, by inquiring how decisions about the city's reconstruction would be made and by whom. I also sought to determine what kinds of assumptions these decisions would articulate about human nature and how the people who made up the complex social landscape of New Orleans would receive them.

The first phase of my work in the summer of 2006 involved attending a variety of recovery-planning meetings that constituted the participatory element of two consecutive—and some would argue competing—planning processes. One was organized by the New Orleans City Council with the assistance of Lambert Advisory LLC and named the New Orleans Neighborhoods Rebuilding Plan (unofficially known as the Lambert Plan). The other was financed by the Rockefeller Foundation with the approval of the Louisiana Recovery Authority and titled "Unified New Orleans Planning" (UNOP). These recovery-planning processes partially overlapped, and their creation was burdened with political power struggles between various levels of city and state government.

What I found most interesting about the Lambert and UNOP planning efforts was that city council members and hired planners heralded them as participatory processes where all city residents, regardless of their socioeconomic background, could play an equal role as authors of New Orleans's reconstruction directive. This, however, was not what occurred in practice. Over the course of my experiences, which are the subject of chapter 5, in these planning activities I routinely documented heated and emotion-laden discussions between urban planners, architects, and New Orleanians who differed in their notions of what it meant to recover from the disaster and what practices were necessary to attain that recovery. Expert planners and architects in these two processes advocated for a conceptualization of the city as a collection of architectural structures meant to move people and replicate capital. People, in these visions of urban recovery, were represented as generic, ahistorical, and subjectivity-lacking beings who could be plugged in and out of different urban contexts and who interacted with one another as strangers, or in what Elizabeth Povinelli (2006) calls stranger sociality.

The Lambert and UNOP planning documents featured images that rendered New Orleans neighborhoods as renovated spaces where buildings featured new, colorful surfaces and sidewalks were adorned with recently planted trees. These same representations also figured people as interchangeable white silhouettes (fig. 3), which moved past cafés, boutique-type stores, and housing structures that planners proposed would replace the large public housing developments of downtown New Orleans. Critics such as Frederic Jameson (1991) and affect theorists like Nigel Thrift (2010) associate such an aesthetic with recent stages of capitalism. I found these images triggered agitated responses from residents of sociopolitically marginalized neighborhoods.

In contrast to the official recovery plans, subaltern city residents insisted that, for them, recovery meant the assisted return of their displaced neighbors, including those who were not homeowners and those who resided in public housing. The New Orleanians argued that their displaced counterparts were not interchangeable people, as the white silhouettes of recovery plans implied. Instead, they saw the displaced as unique and irreplaceable individuals because their sensibilities—the ways they experienced affect—were shaped through life histories particular to the city's neighborhoods. As Cheryl Austin, one of my primary

FIG. 3. Image of New Orleans's possible future from the Neighborhoods Rebuilding Plan. Courtesy of New Orleans City Planning Commission.

interlocutors, remarked one day, "If you did not grow up here, you cannot appreciate living here" (semi-structured interview 2007).

Cheryl and other New Orleanians saw their neighborhoods as more than a collection of architectural and infrastructural features that, when properly arranged and decorated, could turn a profit. Instead, they saw their neighborhoods as spaces that were made and remade on a daily basis through ritual and quotidian practices such as frequenting and operating neighborhood bars, socializing in street spaces, and participating in pedestrian parades known as second lines and Super Sunday.

The recovery-planning processes in question were further complicated because the spaces they were concerned with were not blank slates upon which utopian visions of capitalist development could be rendered. Emotional attachments to specific cultural practices and embodied "New Orleanianness" among the residents of the Tremé neighborhood could not be easily reprogrammed into a desire for neoliberal space. Nevertheless, some gentrifying resident constituencies, local government officials, and planners treated the post-disaster context as a moment in which the city became a new frontier of neoliberal governance, or an open space where relations between public and private sectors could be rearranged and public resources could be directed toward the portfolios and bank accounts of for-profit companies and not the city's most vulnerable residents (see Adams 2013; Button and Oliver-Smith 2008; Johnson 2011).

The case of post-Katrina New Orleans drove home the lesson that space is never a neutral or objective backdrop where social action takes place; instead, space is a product of people's actions and their interactions with the material world (Lefebvre 1992; Low 2011; Raffestin 2007). Keeping these observations in mind and taking a cue from science and technology studies (Latour 1993b; Pickering 2008) and from ecological approaches in anthropology (Biersack 1999; Ingold 2000; Jansen 1998; Paolisso, Gammage, and Casey 1999; Stonich 1993), I began to understand space as something that emerges via co-constitutive interaction between people, their meaning-laden practices, and the materiality of the environments they inhabit. The case of New Orleans illuminated the way people's experience of sensory perception and emotions is cultivated by living in such emergent spaces. This finding urged me to consider how the anthropology of disasters, affect studies, and critical approaches to space and place production could be brought into fruitful conversation in the theorization of the affect's ecology, and that is the subject of chapters 6 and 7.

In chapter 6 I take a look at the devastated New Orleans area known as the Lower Ninth Ward. The chapter traces the interaction of regional development policies, cultural values, and the environment's agency that shaped this urban space and the relationship of this ecology to the variety of ways Ninth Ward residents came to care for the neighborhoods of the Lower Nine and Holy Cross in post-Katrina recovery. Chapter 7 continues chapter 6's line of inquiry but focuses on the reactions of some New Orleans residents and local politicians to the visible—although statistically limited—arrival of Latino and Latin American laborers to work in the city's reconstruction after Katrina. I examine how the presence of this newly arrived population triggered anxieties on the part of some established residents over potential "culture loss." Consequently, the chapter investigates the connections between socially structured space, identity, and sensory experience (i.e., anxiety and comfort) in

New Orleans and situates these affective reactions within the city's deeper history, where people have used spatial distance and quotidian embodiment-forming practices to make and sustain racialized class differences. Finally, the chapter considers how the arrival of Latin American laborers is yet another episode in a long-standing history of migration that has made New Orleans what it is—an ever-becoming creolized urban landscape where emergence and memory collapse into one another in the form of bodies, spaces, and places.

The insights I gained in New Orleans further enriched my understanding of the socio-spatial and affective issues at stake in my preceding research in Honduras and my future work as an ethnographer of post-disaster relocation in southern Illinois and Chiapas. Recently my ethnographic work has taken me to the midwestern United States, where I have followed the relocation process of nearly nine hundred people from the small town of Olive Branch, Illinois, which the Mississippi River flooded in 2011. Chapter 8 uses the case of Olive Branch to discuss the ways community residents and members of a flood mitigation team deal with the attachments people develop to their surrounding social and natural environments, how these various actors imagine what successful relocation should look like, and what possibilities actually exist for materializing the spaces of relocated communities in this part of North America.

Chapter 9 reviews a number of instances in which local government officials, city planners, and gentrifying resident constituencies view post-disaster recovery as an opportune moment to transform their communities, cities, or regions, even if such a change forecloses the possibilities for certain kinds of life for their residents. Chapter 9 uses Foucault's (1978, 1980) concept of biopolitics (the late eighteenth-century transformation in the logic of governance so that the fostering of a population's biological life became a primary concern of the sovereign) and Giorgio Agamben's (1998) theorization of bare life (the conceptualization of the citizen as a life that can be killed but not sacrificed for the sake of the greater biopolitical good) as points of departure in considering how sovereign power is used in post-disaster recovery. When policymakers, government officials, and urban planners experiment with that power, they enter the ethically perilous ground of making decisions about which kinds of lives are worth living and which are not.

Taking Foucault's and Agamben's analyses further, chapter 9 shows how, even though much theorization of the relationship between bare life, biopolitics, and sovereignty focuses on a sovereign power's exclusive privilege to kill biological life, post-disaster reconstruction involves a familiar but different exercise in sovereignty where power is predicated on the right to extinguish certain forms of social life. Using a final case study from Chiapas, Mexico, the chapter engages in a critical analysis of the ethics of biopolitical and neoliberal recovery planning and provides practitioners with analytical strategies for navigating the ethically complex ground of disaster reconstruction. Contrary to recent observations (see Gupta 2012) that disaster contexts are not the most adequate place to examine issues of bare life and biopolitical abandonment, this chapter contributes to a small but growing body of literature (see Marchezini 2015) that highlights disaster reconstruction as fruitful ground for foucauldian analyses and theoretical innovation.

Toward an Affective Approach to Disaster Recovery

The task of this book is not to develop a paradigmatic theory of affect that readily explains the experiences of disaster survivors around the globe, nor is it to grant disaster recovery experts knowledge that is translatable into a power over disaster survivors via claims of knowing them and their circumstances better than they know themselves. Instead, the chapters follow an approach in which the theoretical lessons learned in a particular case create a vantage point from which to ask novel questions in another case and provide methodological threads of inquiry that allow one to devise insights in future research ventures while remaining sensitive to the sociocultural particularities and global interconnections of the places where disasters occur. Mark Thurner (2003) and Dipesh Chakrabarty (2000) have branded similar approaches in social science as "postcolonial," a term they use to denote a move away from Eurocentric metanarratives that assume the history and subjects of Europe are the history and subjects of the world at large.

Together the book's chapters demonstrate that although neoliberal, techno-scientific, and modern governmentality approaches to disaster recovery tend to ignore the sentiments of disaster-affected populations (or figure them as obstacles to be overcome in the name of "rational" disaster management), affect and emotions remain critical dimensions of the ways people experience catastrophes and articulate what it means to recover. *Governing Affect* concludes by stressing the role of sensed experience and emotions in the manifestation of personhood and the complications that emerge when disaster reconstruction programs dismiss or attempt to reconfigure them in neoliberal or modernist terms. Chapter 10 closes by emphasizing that an affective approach toward disasters is by no means superfluous, for conditions such as sociopolitical marginalization or social disruption are ultimately something people feel. An emphasis on affect, in turn, helps researchers and practitioners to address the cultural dimensions of disaster recovery at the nexus of practice, space, and embodiment and to devise aid policies that affected populations find meaningful and relevant.

1. Powerful Feelings

EMOTIONS AND GOVERNMENTALITY

IN DISASTER RESEARCH

It is the summer of 2013, and I have driven two hours from my home in Carbondale to Valmeyer, Illinois, to speak with Lou Ann Simmons, a woman in her early sixties whose house was inundated during the Great Flood of 1993. The Great Flood is considered one of the costliest disasters in U.S. history. The catastrophe resulted from the interaction of human development and settlement patterns in the Midwest during the preceding century, the people's alterations to the hydrology of the region's major river systems, and the unusually high precipitation levels from the fall of 1992 to the summer of 1993. The flood catastrophically affected nine states, caused financial losses estimated at $20 billion, destroyed or damaged five thousand homes, and displaced fifty-four thousand people (Brown, Baker, and Friday 1994).

Following the catastrophe, the majority of Valmeyer's residents relocated their homes out of the Mississippi floodplain to higher ground on the bluffs overlooking the river a little more than a mile away. On this occasion, I am accompanied by a fellow disaster researcher who is conducting a series of interviews with people whose communities have resettled after such events and one of my undergraduate students who is working as an assistant in the project. We are going to talk to Lou Ann because we are interested in why she opted out of the resettlement and chose to remain and live in the floodplain, well within what the U.S. Army Corps of Engineers designated as the hundred-year flood line (an area that has a 1 percent annual chance of flooding). A handful of families have stayed, and rather than being situated within a town of 1,200 people, their houses today are surrounded by grassland and the occasional lonely tree.

Lou Anne's house is a white, two-story Victorian with a porch that has various pieces of wicker furniture arranged for people to sit on and visit. She invites us to join her on the porch, and we begin a conversation about her experiences during the flood and her decision not to resettle. As the sun hangs low on the horizon, Lou Ann brings out personal photo albums and shows us images of the porch we are sitting on under eight feet of water. As we speak, it becomes clear that the flood is a defining moment in her life, not only because of the trauma and material loss she suffered during the event itself, but also because of the dramatic transformation of the social and material landscape of her community that she experiences on a daily basis. One thing that impresses her today is how quiet the floodplain is now that most of her neighbors have moved up to the bluffs. In response to our question of why she stayed, Lou Ann explains that owning and living in a house such as hers was her lifelong dream and that she could not separate herself from it:

> I always wanted a white, two-story house, with a porch. I always wanted to sit outside and see my neighbors go by and talk to them. (structured interview transcription 2013)

Lou Ann makes it clear that, for her, the value of this particular house, with its specific distribution of space (a porch to sit on, two stories), extends well beyond its functionality. She also indicates that the house's worth was once also related to its situation in a web of social relations with her neighbors. The residence has a porch that was, in the past, conducive to having casual conversations with other town residents. The homes built in New Valmeyer (as the relocation community is known), in contrast, do not have porches; they are constructed on slabs and are "right on top of each other" (structured interview transcription 2013). The spatial distribution of the new village is also not conducive to people walking through its streets. The place feels more like a big city suburb, with its winding roads leading to cul-de-sacs rather than following a small-town grid.

Lou Ann explains:

> The new town is nice; it just reminds me of a big subdivision— everything is very similar, nothing is unique about it. Some of the houses down here were two or three stories, just beautiful houses,

and people were scared. I'm sure they had no structural damage; they just didn't want to go through it again. . . . For the ones who I talk to up there, they have regretted leaving. It's not the same as it was down here. I don't regret it at all [staying in the floodplain], not one bit. I love it down here. This is just so quiet, and you get used to it. There's nothing here. There used to be a grocery store in town; the bank, the post office, the hardware—you could walk to whatever you wanted. You get used to it, you learn to make your stops all at one time. (structured interview transcription 2013)

Lou Ann is aware that her house may flood again. She tells us how vigilant she is, keeping an eye on the Mississippi River's level, hoping. She tells us that if her home floods again, that will be it; and she will not come back, for she is too old to rebuild. But until that happens, she will remain living in her dream house, remembering how things used to be when she could sit on her porch and greet her neighbors.

On our drive back to Carbondale, my colleague wants to discuss our conversation with Lou Ann. He brings up the concept of cognitive dissonance, pointing out that what is in Lou Ann's head (her ideal notions of home and community) does not register with the fact that she lives in an area of high flood risk where her own life and that of others (e.g., emergency response personnel who may be called upon to rescue her in case of flooding) may be in danger. My colleague makes a number of allusions to mental health, wondering whether Lou Ann's connection to reality has been compromised by dementia. "How can she not see the danger? Is she crazy?" my colleague asked (fieldnotes 2013).

In response to these questions, I propose that we give Lou Ann the anthropological benefit of the doubt for a moment and that we try not to proceed from the point of departure that we researchers, with our notions of risk and moral responsibility, are right and that Lou Ann, who is engaged in a practice we find irrational, is wrong. What if we were to take what Lou Ann is saying seriously and not dismiss it as cognitive dissonance? What if she cannot make her emotional attachment to her home and to the phantoms of a community that no longer exists (neither on the floodplain nor the bluffs) secondary to the safeguarding of her biological life? What if her corporeally experienced emotional attachment is what makes her life meaningful? What if these emotions and meanings—the things that make her want to live a particular life and run the risk of dying a particular death—are power itself? Should disaster mitigation and risk reduction specialists, for better or for worse, come to terms with people's attachments rather than dismiss them offhandedly as irrationality or "cognitive dissonance"? Most important, how and why did my fellow anthropologist come to think of the people he studies in such a way?

Flood Plain Management: Emotions and
Rationality in Disaster Research

In 1983 Frederick Cuny published the influential book *Disasters and Development*, in which he proposed that disasters could be prevented by applying the techno-scientific governance and engineering practices that make up what he envisioned as development. Cuny specified what he meant by development through the use of two images. He titled the first "Flooding and Its Causes," and he meant it to represent "the poorer developing nations, which we call the Third World" (Cuny 1983, 3). The image is a line drawing of a landscape where far-off hills are labeled "poor farming techniques"; the city, "increased population prevents ground absorption of water"; and farmlands, "flood plains in rural areas attract farmers because of fertile land." The second image, "Flood Plain Management," showcases development in action (fig. 4). The far-off hills are now labeled "contouring of farmland controls water flow, reduces erosion" and "reforestation reduces run-off"; the city, "protective embankments: dikes, walls, etc., help protect urban areas"; and canals, "diversions irrigate farms, channel water into reservoirs" (3).

Cuny's images convey the idea that a standardized collection of technologies and policies could readily harness the agency of the people and geophysical phenomena that shape disasters (e.g., land use and urbanization practices, regional hydrology) irrespective of locality. His figures were idealized landscapes that were simultaneously nowhere in particular and everywhere at once. In the past thirty years, the work of anthropologists and other social scientists who critically examine development practice has raised a number of questions about the assumptions and omissions embedded in such images. Among these assumptions is the

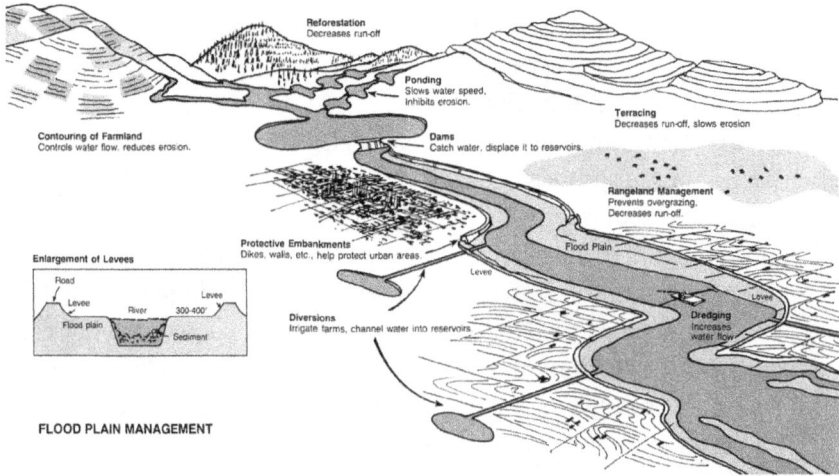

Reforestation
Decreases run-off

Ponding
Slows water speed,
Inhibits erosion.

Terracing
Decreases run-off, slows erosion

Contouring of Farmland
Controls water flow. reduces erosion.

Dams
Catch water, displace it to reservoirs.

Rangeland Management
Prevents overgrazing.
Decreases run-off.

Enlargement of Levees

Road

Levee

River   300-400'

Flood plain

Sediment

Protective Embankments
Dikes, walls, etc., help protect urban areas.

Flood Plain

Levee

Levee

Diversions
Irrigate farms, channel water into reservoirs.

Dredging
Increases
water flow.

FLOOD PLAIN MANAGEMENT

FIG. 4. "Flood Plain Management" from Cuny (1983). Reproduced by permission of Oxford University Press, USA.

idea that the process of social development has a knowable outcome that is uniform across localities. Ethnographic studies (Arce and Long 2000; Ferguson 1999; Holston 1989), in contrast, have shown how development projects always undergo processes of interpretation, contestation, and reconfiguration as people make sense of (or resist) development programs on the basis of their own culturally and locality-specific notions of well-being, social organization (e.g., race, class, gender, and ethnic differentiations), and modalities of political power. Development, rather than something that can be predicted beforehand as "Flood Plain Management" does, is transformed in the moment of practice, and its outcomes are as varied as the people and places where it is put into action.

Another assumption of "Flood Plain Management" is that the civil engineering technologies rendered within its frame will have total stabilizing effects on the materiality of the floodplains where they are applied, creating a static world where "nature" is mastered and where nothing changes over time. More recent anthropological and sociological studies of science and technology (Latour 1993a, 1999; Masco 2006; Mitchell 2002; Pickering 1995, 2008) have demonstrated that such imaginings of technologies at work do not match what actually occurs in practice. Environments—and the ways they react to human intervention—vary significantly from one locality to the next, and people can seldom predict with complete certainty how the material world will respond to their actions and technologies.

In terms of omissions, Cuny's images also erase the colonial histories, impositions, dispossessions, and violences that have given form to the landscapes represented by "Flooding and Its Causes." What is more, they also propose a technological fix for what are, at heart, sociopolitical problems and represent a prime example of development that Ferguson (1994) has called the anti-politics machine. But the omission that concerns me most in this chapter is the absence of people from these images. Certainly one could argue that people are pervasive in these two drawings, even if they are not explicitly rendered, as we see the effects of their actions on the environment. The results of people's actions are present, one could say, in the form of cities, agricultural fields, levees, and terraces that are featured in both drawings.

One could also argue that people are absent as an effect of scale and that understanding the relationship between disasters and development requires pulling away the gaze in a way that makes the individual person irrelevant and emphasizes the impact of collective human action. Still I have returned to the image of "Flood Plain Management" time and time again over my years of disaster research as I have pondered the implications of modernist notions of governance and development for people similar to Lou Ann Simmons who live through catastrophes or reside in places that mitigation experts deem high disaster-risk zones. Whether intentional or not, the emphasis of "Flood Plain Management" on the rendering of nonhumans—be they examples of material culture or environmental features—steers our reflection on disaster mitigation away from people as living, sensing, social, and ecologically emerging beings. Together both images lead their viewers to see their contents as objects, or things in themselves, that must be engaged and related to in emotionally detached and pragmatic ways.

While Cuny's *Disasters and Development* was published three decades ago, it is noteworthy that similar attitudes still pervade in disaster research. In the introductory chapter to the edited volume *The Mangle in Practice* (2008), renowned sociologist of science Andrew Pickering analyzes the condition faced by residents of New Orleans—where anthropogenic coastal erosion and rising levels of the Mississippi River have exacerbated the city's susceptibility to flooding—and states that a key element of addressing the city's vulnerability to disasters involves New Orleanians' letting go of attachment. By "letting go of attachment," Pickering means that the people's impact upon the environment has created circumstances that make life in the city precarious, and to deal with this emergent vulnerability (emergent because such jeopardy did not exist prior to the settlement and growth of the metropolis during the last three hundred years), residents should relinquish their sentimental attachments to this urban area and relocate to a less flood-prone setting.

Like "Flood Plain Management," Pickering makes a pragmatic call for the rearrangement of people and things within a given landscape for the purposes of disaster reduction. While Cuny ignores the affective and emotional dimensions of people's experience of environments and places, Pickering simply recommends their relinquishment. Over the course of my work as a disaster researcher, however, I have come to question the feasibility of such recommendations, and I have also wondered if they gloss over a key dimension of mitigation and recovery. I have learned through my involvement in a number of ethnographic projects that people who live through disasters are socially immersed, living, sensing, and emoting beings, and to represent them otherwise is to miss an important dimension of the experience of being human. After nearly two decades of reflecting on "Flood Plain Management," I am more and more convinced that such representations portray the world as composed of static, unchanging, inanimate objects, and these portrayals limit our ability to understand the affective life—people's embodied, sensed, and emotion-laden relationships to the environments and communities they live in—of the places where disasters occur and the people who reside in them.

I have also learned in my work as a disaster researcher that many people who live in areas that disaster mitigation experts deem high risk often become the subject of moral judgment for their refusal to relocate. In the context of the United States, I have repeatedly heard friends and colleagues utter such question as, Why should my tax money be spent bailing those people out when they knew they were living in a risk zone? What these hasty judgments miss is that, in many instances, people such as Lou Ann Simmons come into worlds that have a long history and their creation is as much a result of the practices (e.g., settlement, home construction, agricultural production, trade) of people in the immediate vicinity of disaster-prone areas as well as those of others who, although geographically and temporally removed, also play a role (e.g., through development policy and practice at national and international levels) in shaping the landscapes where disasters manifest. What is more, as social scientists who work on disasters have demonstrated since the 1980s, those who are most vulnerable to the materially destructive and socially disruptive effects of geophysical phenomena often face great challenges in their social and spatial mobility. Granted, even the most economically marginalized people have tremendous agency and are capable of migrating and creating new communities in the most socially inhospitable conditions imaginable, but doing so on their own meaningful and affectively familiar terms is another story. There is much work to be done in this regard (Marino and Lazrus 2015).

Ecologies of affect are also political ecologies (Biersack 1999), and I understand the political ecology of affect as the multidimensional co-constitutive relationships between policy, technology, imaginaries of social development, human practices, and the agency of the material world, where environments, sensing bodies, and risk itself come into being. To pass judgment on individuals like Lou Ann is to err on the side of a reductive victim-blaming logic and to fail to understand the more complex and not readily apparent processes that engender catastrophes. Let us also not forget the close connection between subalternity, environmental injustice, and disaster vulnerability (Button 2010; Hoffman and Oliver-Smith 1999). If we do not engage this complexity, we fall short of understanding the relationship between attachment and disaster risk in a distinctly anthropological way.

In this chapter, I lay out the cultural and historical background that made images such as "Flood Plain Management" and Pickering's statement possible, and by doing so, I bring into question their status as self-evidently rational ways of speaking and thinking about risk and disasters. I pay specific attention to the ways emotions have been figured throughout the history of modern ways of thinking, governing, knowledge making, and technology production. This history of emotions and modernity serves as a foundation for the following chapters, where I contrast the ways emotions came to be figured in modern thought and the ways that people who live through disasters use emotions to make sense of their experiences, assess the relevance and usefulness of disaster mitigation programs, and reflect on what it means to recover from a catastrophe.

Governance, Modernity, and Emotions

Images such as "Flood Plain Management" express desires that are rooted in specific social contexts and have unique histories. Cuny's (1983) figure expresses a longing for a societal state of affairs where dispassionate experts make governance decisions that enable an optimal techno-scientifically crafted balance between human and nonhuman agency that renders the world stable. These decisions are not led astray by political interests or individual passions. Cuny's desire for a rule of dispassionate experts was not entirely novel in the late twentieth century when *Disasters and Development* was published, and it may very well have had an evolving history extending over two millennia.

In the fourth century BCE, Plato recorded a kindred vision of sociopolitical organization in *The Republic* (2003 [380 BCE]), where he imagined a society in which ascetic philosophers who were disconnected from anything that may have aroused feelings of attachment or personal interest (e.g., kinship) governed on the basis of rubrics that were conducive to the achievement of what he considered civilization's pinnacle. For Plato, such rubrics were accessible to philosophers through practices like geometric proportionality, which allowed people to discover what he viewed as the perfect social order. In this case, elements of the built environment like the Athenian Parthenon were thought to be expressions of geometrical balance and a metaphor for the ideal republic. A perfect society was one of perfect proportions. Politicians, Plato insisted, could not be trusted to preside over societies, as they were prone to using force and creating inequity in the search for personal gain. In the Mediterranean world and Europe, Plato's ruminations on governance, ethics, and society are part of a lengthy and intricate history of attempts to organize societies on the basis of principles derived from systems of order that are thought to be independent of, to preexist, and to transcend social orders (i.e., geometry, nature). This lengthy history is also the unfolding of a perspective on knowledge making and social organization that has gone by the name of modernity.

Although Plato's utopian republic never came to pass, two thousand years later, Robert Boyle, the person credited with the invention of the scientific method, found himself pursuing a related query: how can people ascertain universal and unbiased truths about the operations of the natural world—matters of fact—upon which decisions about the governance of society could be based? Boyle's invention of the scientific method emerged as a response to the social upheaval of the seventeenth-century English Restoration, a time when people in England and Europe were questioning the founding principles of feudal sovereignty (Shapin and Shaffer 1985). Boyle's intention behind the scientific method was to devise observation and documentation techniques that allowed people to see, agree upon, speak about, and disseminate knowledge unencumbered by cultural biases (Bauman and Briggs 2003). The knowledge derived through these observation and documentation techniques became matters of fact, which practitioners of the scientific method considered true anywhere and at any time.

The development of the scientific method had significant implications for the ways modernist thinkers viewed affect and emotions. Boyle and his supporters expected the maker of scientific knowledge to embody the subject position of a dispassionate observer, or a modest witness, and they considered a propensity toward emotional excitement an impediment to objectivity and rationality. Given the cultural conventions concerning gender, class, and race of the time, advocates of the scientific method considered women, the working classes, and racialized others (e.g., people from equatorial latitudes, colonial territories, and southern Europe) to be unsuitable for scientific work, for they considered these others as prone to extreme emotional states that hindered their ability to document matters of fact (Haraway 1997). While Boyle envisioned the laboratory as a space where a culture of no culture could manifest, seventeenth-century scientists brought culture into the laboratory in the form of racist, sexist, and classist prejudices concerning who could be a scientist, as well as the upheaval of the Restoration, which set a sociopolitical condition as the reason for scientific practice.

For historians and philosophers of science, Boyle's creation of the scientific method marked a watershed moment in the cultural history of western Europe, the moment when the claim to a particular way of seeing and speaking about the world as exemplified by "Flood Plain Management" came into being—that is, the modern ways of knowing and knowledge making. To claim epistemological modernity (objectivity) was to claim having an ability to tease out what is social from what is natural, or to separate objects from subjects, and therefore have an exclusive access to the truths and facts of the world (Latour 1993b). The rise of epistemological modernity was also characterized by attempts to devise ways of speaking and writing that were considered conduits of objectivity and that differed from the ways subaltern populations (e.g., women, colonial others, working classes) used language (Bauman and Briggs 2003; Briggs 2004).

Claims to objectivity were exactly that, however; they were not actually achieved in practice (Haraway 1997; Latour 1993b). On the contrary, if there is one thing techno-science has facilitated, it is the proliferation of intimate fusions between culturally situated values and materiality through laboratory practice while claiming not to do so, and therein lies the power of modernity: one can claim not to do what non-moderns do even as one does it (Latour 1993b, 1999). What is of interest here, however, is how Boyle came to figure emotions as an impediment to reason. Emotions had to be put aside and kept out of the process that would yield the matters of fact upon which decisions about the organization of political power and society could be made.

Boyle was not alone in his indictment of emotions as obstacles to reason and objective knowledge. In *The Metaphysics of Morals*, Immanuel Kant also represented emotions as inhibiting a person's ability to be ethical, with ethics, in this case, being "self-constraint in accordance with (moral) laws" (1996 [1797], 186). For Kant, people who pursued actions such as revenge on the basis of long-held anger were incapable of such self-control and were therefore immoral, whereas those who could put their emotions aside in making decisions were virtuous. Additionally, rash actions based on emotion and taken in the heat of the moment were not as reprehensible in Kant's eyes as those that required stewing in an emotional state for a prolonged time. Emotions, and affect in particular, were therefore figured as base instinctual reactions, naturalized as if part of a state approximating nature, and thus were distant from civilization. In Kant's own words: "Man has a duty to cultivate the crude predispositions of his nature, by which the animal is first raised into man" (195). It is also noteworthy that Kant specified an analytical language for reflecting on emotions by defining a difference between affects and passions:

> *Affects* and *passions* are essentially different from each other. Affects belong to *feeling* insofar as, preceding reflection, it makes this impossible or more difficult. Hence an affect is called *precipitate* or *rash* (*animus praeceps*), and reason says, through the concept of virtue, that one should *get hold of* oneself. Yet this weakness in the use of one's understanding coupled with the strength of one's emotions is only a *lack of virtue* and, as it were, something childish and weak, which can indeed coexist with the best will. It even has one good thing about it: that this tempest quickly subsides. Accordingly a propensity to an affect (e.g., *anger*) does not enter into kinship with vice so readily as does a passion. A *passion* is a sensible desire that has become a lasting inclination (e.g., *hatred*, as opposed to anger). The calm with which one gives oneself up to it permits reflection and allows the mind to form principles upon it and so, if inclination lights upon something contrary to the law, to brood upon it, to get it rooted deeply, and so to take up what is evil (as something premeditated) into its maxim. And the evil is then *properly* evil, that is, a true *vice*. (208, italics in original)

According to Kant (1996, 208), then, affects "belong to feeling" and differ from passions by "preceding reflection," while passions are "a sensible desire that has become a lasting inclination." He therefore saw a distinction between hatred, which he understood as a passion, and anger, which he considered an affect, or a rash emotional reaction. Passions, moreover, led to evil in cases in which "inclination lights upon something contrary to the law." Kant's differing moral valuation of affects and passions reflects a cultural convention that still plays out today as an unquestionable truth. For instance, in the case of the U.S. legal system, premeditation carries a higher penalty for first-degree murder than does unplanned manslaughter carried out "in the heat of the moment." In *The Metaphysics of Morals*, Kant also makes a connection between his appraisal of affects and passions and governance, although he is concerned with the governance of the self, not a state. He writes: "Since virtue is based on inner freedom, it contains a positive command to a man, namely to bring all his capacities and inclinations under his (reason's) control and so to rule over himself, which goes beyond forbidding him to let himself be governed by his feelings and inclinations (the duty of apathy); for unless reason holds the reins of government in his own hands, man's feelings and inclinations play the master over him" (1996, 208).

As Kant brings up the notion of "the duty of apathy," he makes a proposition similar to Boyle's "modest witness" by prescribing "moral apathy" as a disposition from which to make rational decisions. He writes: "The word 'apathy' has fallen into disrepute, as if it meant lack of feeling and so subjective indifference with respect to objects of choice; it is taken for weakness. This misunderstanding can be prevented by giving the name '*moral apathy*' to that absence of affects which is to be distinguished from indifference, because in cases of moral apathy feelings arising from sensible impressions lose their influence on moral feeling only because respect for the law is more powerful than all such feelings together" (1996, 209).

Kant would have considered Lou Ann Simmons to be a person governed by her affects, incapable of taking the subject position of "moral apathy" or the "modest witness." While Kant and Boyle believed that their figuration of emotions, affects, and passions as obstacles to reason and virtue was a proclamation of a universal truth, Catherine Lutz (1986) understands such a take on emotions as a culturally situated perspective on sensory and emotional experience. Similarly, Webb Keane (2007), Richard Bauman and Charles Briggs (2003), and Donna Haraway (1997) have interpreted the rise of modern ways of knowing (e.g., "objectivity," positivism, the scientific method) as a process through which the subjectivities of European male intellectual elites came to be represented as objective ways of relating to the world and people.

The implications of the cultural horizon that made Kant's ideas about emotions and rationality possible for governance can be recognized in cases such as that of colonial India, where English administrators deemed local subaltern populations incapable of self-rule given their cultural conventions concerning emotion. Colonized populations in South Asia, for example, did not consider public displays of emotion such as crying (especially when recognizing the suffering of another) an impediment to rationality. Instead, South Asians considered such emotional displays to be manifestations of a divine capacity to recognize another's pain (Chakrabarty 2000). Not unlike Kant, English colonizers judged these emotional reactions as proof that their colonial subalterns were not in proper control of their affective experiences, were incapable of rationality and virtue, and were therefore unfit to govern themselves. Moreover, Edward Said (1979) has identified the characterization of subaltern people and their actions as emotional (and therefore "irrational") as a well-worn strategy of colonial and hegemonic elites that allows the latter to dismiss the former's affect-laden acts of resistance as lacking in rational political legitimacy.

Presenting an alternative perspective on what Kant would call passions, Didier Fassin (2013) has made the case that emotional states like resentment must be reevaluated in contextualized and nuanced ways. In some instances, as in the cases of the Holocaust in World War II and apartheid in South Africa, resentment can become an important element of collective memory; its experience is necessary to ensure previous atrocities are not repeated and to remind us that some inequities and violences remain without atonement. Fassin also argues that the experience of affect and emotion must be understood from the vantage point a person occupies (or is relegated to) in a particular body politic. Referencing Friedrich Nietzsche, Fassin stresses that the noble person is just that—a noble who does not bear resentment due to his or her privileged position.

Boyle's and Kant's quests for objectivity and rationality also took place within a broader transition in the ways people legitimized sovereign power, imagined the relationship and obligations between the sovereign and their subjects, and thought about the reasons for government. For Michel Foucault (1978, 1991), this reflected a transition between notions of governance: in the sixteenth century the sovereign's object of concern was keeping and expanding his or her estate, but in the late eighteenth century the purpose of sovereign power became caring for and fostering human populations as a collection of biological organisms. In the former, the sovereign's concern with the well-being of the people who lived within his or her estate (which was envisioned as an extension of the sovereign's body) went only so far as the people were seen as an extension of that estate. In this case, an assault on the estate and its people constituted an assault against the metaphorical body of the sovereign and was punishable by death. In the latter, sovereign power was charged with (and legitimized by) the obligation of caring for human populations as biological beings; with this care emerged a collection of sciences and techniques of population management including public health, modern urbanism, and economics. It is not at all inconsequential that these sciences and techniques are now part of the tool kit for many NGO program managers and disaster recovery planners.

While some commentators on the political history of western Europe may see such a transition as a progressive move toward an improved condition, Foucault simply saw it as a transformation of the way power operated and was justified. This new organization of power, in turn, produced a new collection of ethical complications, ways of thinking about and experiencing the self, and forms of violence. Foucault named this novel modality of power "biopolitics" and the sciences and techniques of biopolitical governance "governmentality." Hence, as illustrated by this chapter's initial anecdote, flood mitigation specialists and anthropologists became preoccupied with Lou Ann Simmons's decision to remain living in the Mississippi River floodplain and dismissed other ways of imagining ethics and the good life.

In terms of affect studies, what is interesting in this shift is that while the quest for epistemological modernity exemplified by Boyle's and Kant's work involved a claim to the suppression of emotion, the emergence of modern governmentality and biopolitics did not so much involve the abolishment of emotions as it did their careful recultivation and repatterning. In the context of biopolitical governance, it involves a fostering of desire, attachment, and affection for the body as a biological entity, as well as the cultural practices that cultivate a biopolitically optimized person (Povinelli 2006; Stoler 1995). From this perspective, power becomes instantiated in multiple mundane contexts as people engage in practices that structure the experience of affects and emotions such as terror, disgust, and desire in relation to specific culturally marked bodies and embodiment-making habits.

Rather than operating through processes or rational choice, where modest witnesses and moral apathy suppress emotions in the name of ethical decision making, systems of modern governance operate through the inculcation and pursuit of biopolitical affections (e.g., fear the terrorist, be disgusted by the HIV-infected person, judge the woman who chooses to remain living in the floodplain). In the case of disaster mitigation, these affections are exemplified in Cuny's (1983) "Flood Plain Management," where the preservation of biological life and the prevention of economic loss through flood protection systems and settlement patterns that remove people from floodplains are given priority over the attachments people may have to a particular place and the socialities in which they are involved. As I have learned over the course of my ethnographic experiences, in the imagination that makes "Flood Plain Management" possible, the people who live in flood-prone areas are deemed irrational and, at times, outside the bounds of legality and morality for potentially putting their lives and those of emergency first responders at risk. We meet such people in the chapters that follow and see how their attachments to place and social relations both challenge biopolitical approaches to disaster mitigation and call for an understanding of their experiences and priorities in affective ecological terms.

A Historical Ecological Approach to the Anthropology of Affect and Emotions

Up to this point, I have relied on a number of terms customarily used to describe people's sensory experience of the body to set up this book's thematic core. I have used terms such as *affect, emotion, sentiments*, and *sensory perception* to recognize a point of tension in the encounters between disaster-affected populations, government agencies, and NGOs. The meaning of these terms and the experiences they are supposed to apprehend, however, are by no means standardized in the anthropological literature. In some instances, affect and emotions are used interchangeably (Hsu 1977). In other instances, affect scholars prefer to establish key distinctions in their use of the two terms, as in the case of Brian Massumi's essay "The Autonomy of Affect" (2002). Basing his observations on the work of neurological science, Massumi's take on affect is that people's bodies (and, more specifically, their neurological systems) constantly experience multiple and simultaneous sensory inputs, but these inputs do not necessarily become something people recognize, name, and speak about—that is, an emotion. When sensory inputs and reactions become a subject of thought and narrative, Massumi believes they cross a qualitative threshold and become emotions. In Massumi's (2000, 88) words: "An emotion is a subjective content. . . . Emotion is qualified, the conventional, consensual point of insertion of intensity into semantically and semiotically formed progressions, into narrativizable action-reaction circuits, into function and meaning. It is intensity owned and recognized. It is crucial to theorize the difference between affect and emotion. If some have the impression that it has waned, it is because affect is unqualified. As such, it is not ownable or recognizable, and is thus resistant to critique."

To explain the difference between affect and emotion, Massumi uses the example of former U.S. president Ronald Reagan and his fame as the Great Communicator. Reagan's electoral popularity, Massumi argues, did not lie on the logic of the former president's spoken words. Instead, Reagan's capacity to inspire and attract voters rested on the ways he used his body language and the tonality of his voice to make people *feel* as though he was a superior political candidate; in this way he embodied a particular kind of white masculinity (although Massumi falls short of directly addressing issues of race and gender in his analysis). Massumi considers Reagan's electoral popularity as having operated at an affective level because it was not so much the content of his speeches that drove people to the polls in his support as it was the way his voice and culturally cultivated body language evoked sensory reactions from voters. These sensory reactions remained "resistant to critique" because people could not put into words the actual reasons why they preferred Reagan (racialized and gendered affective preferences that eluded reflection), and when they did, they misleadingly attributed their preference to his verbal communicative skills. This power to move people for reasons they cannot discursively articulate, Massumi argues, is affect. Emotion, in contrast, is "intensity owned and recognized," meaning that it shares a kindredness with affect in terms of sensory perception but lies at the opposite end of a continuum of consciousness where it is recognized, named, and made personal. Affect, for Massumi, is that which we feel but cannot put into words, while emotion is affective experience that has been brought into the social realm through language and therefore named and considered carefully.

Massumi's reflections on affect and emotions, however, have not gone without criticism. Recently Emily Martin (2013) has questioned Massumi's reliance and uncritical acceptance of neurological science as a point of departure in his development of affect theory. Martin rightly claims such reliance gives credence to the tendency of neurological science to represent its knowledge as an access point to a biology that is allegedly independent of human behavior and culture and is universal. In doing this, Martin argues, Massumi misses the importance of recognizing how subjectivity comes into being via cultural practice and what the role of subjectivity is in the experience of the senses and emotions.

Although Martin's critique is important, I think it falls short of realizing the potential of an affective focus in anthropology. By emphasizing the role of subjectivity, Martin reinstates the modern epistemological gaze that claims to divide the world into a collection of dichotomous oppositions—for example, objects versus subjects and nature versus culture—but a number of anthropologists and social theorists recognize such dichotomies as being more of a culturally particular conceit of modernist thought than an objective understanding of how the world at large works (Latour 1993b). A focus on affect is promising because it requires us to understand the relationship between cultural practice and the materiality of bodies in the making of embodiment, and this relationship cannot be understood through a neat separation of subjects from objects and subjectivity from objectivity.

One of the questions Massumi's essay and Martin's critique provoke is, if affect is an unqualified intensity, by what body is it felt? I ask this question in light of a number of anthropological studies that claim the body is by no means a universal and purely biological object; rather, the body is something that comes into being at the intersection of cultural practice (power) and materiality (the body's "biology"). Judith Farquhar, for example, argues that the body in Chinese medicine is "a flavorful temporal formation" (2002, 75), meaning that bodies that react to the flavors and scents of herbal remedies (affective impacts) in specific ways are shaped, over time, through the practice of this healing system. Susan Bordo (1993), in turn, has insisted that those bodily practices that seem most natural (and therefore biological and universal) are manifestations of the ways cultural categories, values, and norms become embodied through everyday actions.

Elizabeth Povinelli (2006) has also weighed in on this issue, urging anthropologists to recognize how the experience of love—which, for Massumi, would fall more within the realm of emotion—is borne at a bodily level within a particular societal context that involves the dense entanglements of the body's sensing capacities, quotidian practice, and political economy. We learn to experience love, Povinelli argues, in distinct ways within the social and built environments of cultural contexts like economic liberalism. Similarly, Lutz (1986, 1988) and Lila Abu-Lughod and Lutz (1990) encourage us to recognize what they call the "cultural construction of emotion" and to reflect on the ways people in varying social settings differ in how they parcel out the emotional world: what an emotion is called and how it is felt in relation to a thing or a person.

Even in neurological terms, Claudia Castañeda's (2002) work on childhood development has shown that the human body is not a standardized, universal biological entity and that the composition of children's brains takes shape in relation to culturally contingent child-rearing practices. The body that senses, then, is dependent on the socio-environmental milieu in which it comes into being, and affect, although an unqualified intensity, is mediated through these ecologically situated bodies and is therefore not, as Massumi verges on suggesting, pre-social.

Tim Ingold (2000), for his part, also reminds us that organisms (bodies) are not made outside of an environment (both social and "natural") but are shaped in relationships with humans and nonhumans; therefore, they are unique to the ecological circumstances in which they take form. In referencing ecology, in this case, Ingold means the co-constitutive and meaning-laden relationships among people and between people and the things a person experiences, is shaped by, and modifies to fit changing socio-environmental circumstances in a particular locality. This use of ecology, which I adhere to in this text, differs from uses that convey the idea that human ecologies are bounded, coherent systems that either maintain balance through internal mechanisms or are brought back into balance by human action (e.g., ritual; see Rappaport 1984). My use of ecology, in contrast, is more in line with alternative uses of the word (see Ong 2005), where it conveys the idea of mutually shaping and ever-shifting, open-ended, and emergent relationships between people, the products of cultural practice (e.g., technologies, policy, values), and materiality. Additionally, these relationships are not necessarily stable, monistic, or harmonious.

What is particularly interesting about disasters is how the extensive disruption to the ecologies where bodies, affect, and emotions manifest generates moments when disaster survivors reflect on what—for Massumi—is unqualified intensity; that is, they think about and bring into discourse that which ordinarily is sensed but not qualified or reflected upon. In *Governing Affect*, we see disaster survivors who struggle to put into words their bodily and sensory experiences of recovery (the impressions that devastated or reconstructed environments make upon their bodies) and to convey to recovery planners and NGO program managers how they define reconstruction in affective terms—that is, the materialization of ecologies that allow them to experience a sense of well-being, familiarity, and comfort.

Ethics, Power, and Practice in Disaster Research

The relevance of the preceding observations on governmentality and affect to disaster research is that in the early twenty-first century, the governmental and nongovernmental agencies charged with the task of mitigating or responding to disasters are all, in one way or another, connected in their techniques, objectives, legitimizations, or missions to biopolitical understandings of governance and well-being. In the following chapters, we see how techniques of recovery project management and urban design, whose histories intertwine with the emergence of governmentality, manifest in the aid practices of governmental organizations and NGOs, and we see how these techniques come into tension with the affective experience of recovery, place, and community of many disaster survivors. At the same time, we see how these practices of recovery assistance on the part of governments and NGOS are by no means identical and how they change as people interpret and reconfigure them from one disaster-affected locality to the next. This is particularly the case in postcolonial contexts, where biopolitical logics and practices become entangled with location-specific social orders of race, class, and gender, mutating and multiplying manifestations of neoliberalism and modernity.

The foregrounding of a concern with affect and emotions, however, should not be taken as a call for disaster mitigation experts and governmental officials to uncritically accept the sensory perceptions and feelings of disaster survivors as a roadmap for recovery. The cases reviewed in subsequent chapters show how historically disenfranchised and subaltern populations often speak of social justice and cultural rights in affective terms. Marginalization, exclusion, and inequity engendered through reconstruction programs are things they *feel*. At the same time, though, disaster-affected populations are not lacking in complexity, social hierarchies, and multiple levels of subalternity (e.g., class, gender, racialized differentiations). The complexity and inequities that exist among disaster-affected populations are often instituted and sustained at an affective level as well, as when local elite groups impose what they find pleasant, comforting, and nostalgic in the practice of disaster recovery.

It is also worth keeping in mind that the political ecology of disasters has the effect of inequitably distributing a catastrophe's socio-material impacts along the fault lines of a society's body politic, often putting socioeconomically marginalized sectors of an affected population at greater risk than others. This book's affective focus, then, brings social and environmental justice concerns into articulation with the epistemological issues surrounding emotions in modernist thought—that is, the representation of politically disenfranchised disaster survivors as emotional and therefore irrational beings incapable of self-governance. Asking whose emotions and attachments are disregarded and whose are prioritized in disaster contexts also helps us recognize emotions and affect as politically active positionalities in contexts of socioeconomic inequity (I am grateful to one of two blind peer reviewers for this particular insight).

Finally, by proposing an affect- and emotion-centered approach to disaster recovery, I do not advocate for a simplistic documenting of disaster survivors' sensibilities and dispositions. Rather, I propose that affect and emotion be taken as a space of critical examination and dialogue between assisting governments, aid agencies, and disaster survivors where reconstruction and mitigation projects that are meaningful to disaster survivors can be devised. Rather than proposing ready-made solutions to the affective challenges of recovery and mitigation, the purpose of this book is to assist its readers in becoming keen analysts of affective issues who are capable of devising aid policies and practices in collaboration with those who live through catastrophes.

2. *Hallarse*

DEFINING RECOVERY IN AFFECTIVE TERMS

Following the devastation triggered by Hurricane Mitch in October 1998, 904 flood-displaced families resettled from the city of Choluteca, Honduras, to Limón de la Cerca, the largest post-disaster resettlement site in the country's southern region. At the time of the disaster, Choluteca proper was what the Honduran National Institute of Statistics called a "medium sized city" of 70,000 people and located within a municipality known by the same name whose population was estimated at 120,000. When I arrived to conduct my ethnographic research in 2000, Choluteca was an interesting urbanization phenomenon. A small town with mining and agricultural roots extending to the pre-Columbian period that had grown in response to its involvement in transnational circuits of cattle ranching, aquaculture, and melon production and exportation, I found Choluteca difficult to describe. At that time, the city was a mix of high-traffic boulevards, where exhaust-spewing semitrucks moved commercial goods in, through, and out of the urban area; small stores; and quiet residential neighborhoods of various socioeconomic levels.

Cholutecans themselves had an ambivalent relationship to their hometown. On one occasion, during a conversation at a *pupusa* (a grilled corn patty stuffed with cheese and chorizo) restaurant, I remember Myrna Portillo, a young Cholutecan I came to know in my work, commenting, "Esto no es una ciudad, es un pueblón" (This is not a city, it's a *pueblón*) (unstructured interview 2001). *Pueblón* is a term used in Central America to describe a settlement that has outgrown itself, where urban sprawl combines with provincial life. The specific site of my research, however, was located seven kilometers away from Choluteca along the Pan-American Highway, a major international road that connects North, Central, and South America (fig. 5).

The construction of Limón de la Cerca (referred to as Limón in the

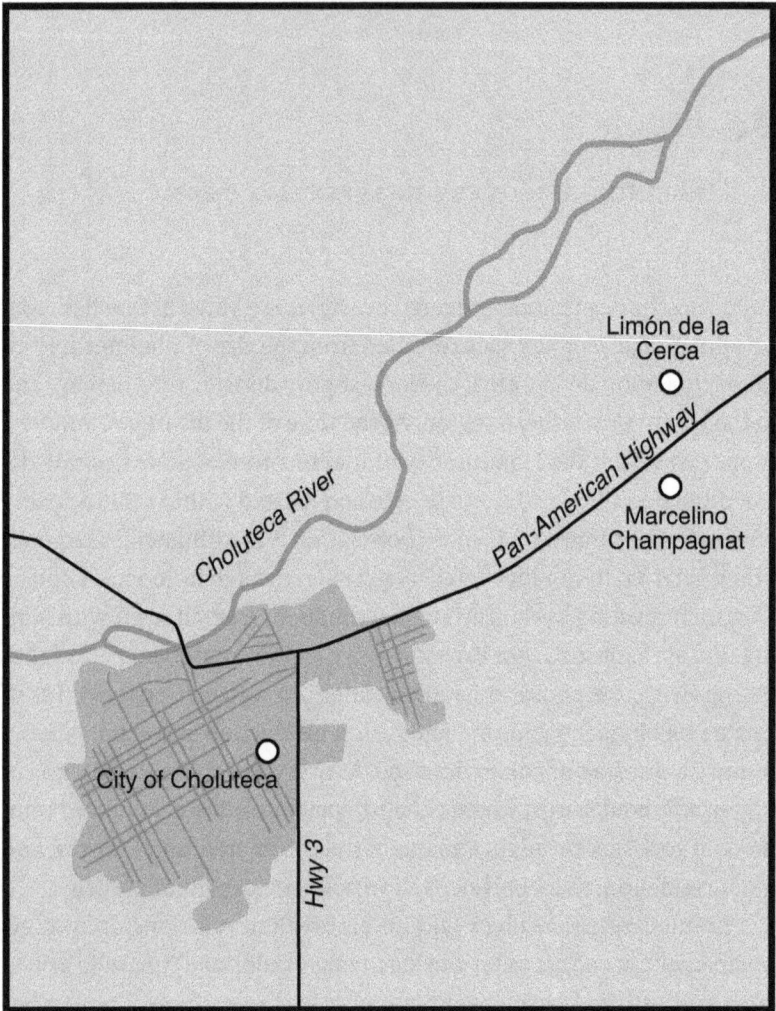

FIG. 5. Map of greater Choluteca urban area. Courtesy of author.

rest of the text) began in January 1999, and its various infrastructure and housing projects were funded by a number of foreign governments and private companies, including the U.S. Agency for International Development and Samaritan's Purse (a United States–based evangelical NGO that built more than half of Limón's twelve hundred homes). I first came to know this resettlement site in July 1999, when I traveled to Honduras to collect growth measures of children younger than five years of age as part of a nutritional health study. Local public health officials directed me to the Limón, indicating it was the best location to find hurricane survivors to include in the survey. At that time, residents and assisting NGOs were still in the process of constructing the community's new housing units. It was the rainy season, and muddy streets and pools of stagnant water surrounded the half-constructed homes. Honduran public health workers had documented cases of hemorrhagic dengue fever and malaria, and the majority of families did the best they could to weather the site's harsh conditions while living in diminutive temporary quarters called micro-shelters or micros (fig. 6).

Residents of Limón were not the only disaster-displaced Cholutecans in the area. Across the Pan-American Highway, a few hundred feet away, a smaller group of approximately six hundred families had also resettled and formed their own nascent community, which (for reasons I will later explain) they christened Marcelino Champagnat (referred to as Marcelino in the rest of the text). Conditions in this second site seemed even more precarious than those of Limón because none of the families had micros; they all were living in canvas tents. At that time, the harsh rainy season's conditions made me wonder if they would survive another six months.

My first visit to Choluteca in 1999 was brief, and I did not return to Honduras until June 2000, when I arrived to conduct a thirteen-month ethnography of the resettlement process. Visually, Limón had changed significantly in the year that had passed. Nongovernmental organizations involved in the recovery and displaced Cholutecans had completed the bulk of the home construction projects, and only a few families remained in micros. At the center of the community, aid organizations had also built a health center and a primary school.

Still, some things gave me pause as I reacquainted myself with the resettlement locale. Of the 1,200 constructed homes, 300 remained vacant, their owners preferring to rent rooms in Choluteca rather than live in Limón. Key infrastructure projects, like the community's electrification, had not been completed despite ample donor funding from foreign governments. Housing structures also seemed to be inadequately constructed to suit local environmental and social conditions. Haphazardly attached tin roofs (rather than the customary ceramic tiles) made for suffocatingly hot interiors and were known to blow off during heavy

FIG. 6. A micro-shelter (white structure) distributed by the International Organization for Migration, a wind-damaged house constructed by Samaritan's Purse, and electricity poles without high-power cables. Courtesy of author.

FIG. 7. Mara Salvatrucha graffiti. Courtesy of author.

thunderstorms, resulting in deaths and severe injuries. Housing structures were also diminutive by local (and perhaps any) standards, featuring single-room, 270-square-foot designs. The houses were located on 1,300-square-foot land parcels that left little room for future expansions, making resettled Cholutecans feel constrained. At the same time, municipality guidelines prohibited residents from acquiring more than one land parcel. The walls of many houses also bore graffiti identifying the community as the territory of the Mara Salvatrucha, one of the two major transnational street gangs operating in Honduras at the time, which I discuss more thoroughly in chapter 4 (fig. 7). In the short time of its existence, Limón had gained a reputation among city officials and Cholutecans as a place of violence and delinquency. My conversations with Limón residents also informed me that *mara* (gang) members operated with impunity in the resettlement site.

Across the Pan-American Highway, in contrast, I observed very different and unexpected conditions in Marcelino. The people I thought might die of malaria and dengue within six months in 1999 seemed to be thriving. Working together, aid organizations and displaced Cholutecans had replaced the canvas tents with homes that were 60 percent larger than those in Limón. These housing structures had the internal partitions to separate living and sleeping quarters that Cholutecans were accustomed to. The houses also sat on larger land parcels that were twice the size of those in Limón and accommodated the residents' planting vegetable gardens and fruit trees and raising pigs and chickens—both of which were customary among the displaced prior to the storm. The community also distinguished itself through the absence of street-gang graffiti, and the *maras*, although present, were subordinated to an informal network of community organizers who subjected gang members to weapons searches and issued ultimatums to those *mareros* (gang members) known for excessive acts of violence. Finally, NGOS and Marcelino residents collaborated on completing infrastructure projects in a timely manner, and the latter boasted having "the best electrification project in all Southern Honduras" (fieldnotes 2000).

As an ethnographer of disaster recovery, I was interested both in learning how the displaced Cholutecans themselves experienced the conditions taking shape in both communities and in understanding why the two communities had taken such different attributes. What I found over the course of my ethnographic work was that Limón residents routinely called attention to the differences between the two communities and used affective terms in speaking about their experience of the social, spatial, and material conditions of their reconstruction site. Furthermore, this ethnographic research revealed that the differences between the two communities were the result of the relationships between disaster survivors, NGO staff, and local government that occurred during the recovery process; these relationships were mediated through postcolonial practices of political culture and modernist techniques of disaster recovery management. In this chapter, I use the comparison of Limón de la Cerca and Marcelino Champagnat to highlight the relationships between aid organizations, government officials, and Cholutecans that either created an affective experience of recovery or perpetuated the undesired socio-material impacts of disaster among the latter.

Speaking about Recovery in Affective Terms

On a bright and hot August day in 2000, I sat down with Doña Julia Maradiaga, a resident of Limón, to talk about her assessment of life in the resettlement site. A woman in her mid-forties, she shared a house with her young adult daughter and two grandchildren. When I asked how life in Limón compared to her life in Choluteca before Mitch, Doña Julia replied, "Digamos que yo, aquí, yo no me hallo" (Let's say that I, here, I don't *hallo* myself) (structured interview transcription 2000). I found Doña Julia's choice of words interesting. The word *hallo* is a conjugation of the Spanish verb *hallarse* (to find oneself). Explaining why she could not "find herself," Doña Julia continued:

> We don't have water, we don't have light. Sometimes I have to sit outside because these houses are so hot. Look at the houses of Marcelino, how pretty, because over there they have [internal] divisions. They feel fresh and there is more space. And here we have large families, you know, everything is reduced. I also see that my house is cracking, and I sit here and think that it's going to fall down. It's cracked all the way from the top [of the wall]. It's cracked there by the window, and one thinks, how many strong winds will the house be able to stand? Over there, two houses went down. (structured interview transcription 2000)

At this point in our conversation, Doña Julia's neighbor, who overheard us talking, chimed in: "Where was it that one of the rooftops caved in and it killed two children?" Doña Julia continued:

> Over there, by La Samaritana [a neighboring subdivision of Limón]. Over there two children died because the wind lifted one of the rooftops and one of the walls collapsed, and it fell on top of them. (structured interview transcription 2000)

In this chapter, I call attention to the way Doña Julia sums up her assessment of life in the reconstruction site through the use of the Honduran Spanish colloquialism *hallarse*. The Spanish verb *hallar* indicates the act of finding something, but in Choluteca, it is also used as an adjective accompanied by the verb *sentir* (to feel). The phrase *no me siento hallada* literally translates into English as "I don't feel myself found." For Cholutecans, however, *hallarse* denoted a sense of finding oneself in a familiar and comforting environment that was built and social at once.

Like Doña Julia, other Limón residents also invoked *hallarse* when responding to my question about their lives in the resettlement site. On one occasion, Don Julio Rodriguez, a middle-aged displaced Cholutecan, assessed the outcome of reconstruction projects:

> Well, we lived better in Choluteca. We had our house, and that's where we were born and raised, in Barrio La Cruz. But since we had to come to look for shelter here, without wanting to . . . Look at that bandit Hurricane—it took our houses and everything, and it left us in the street. And, well, it's that here I don't, here I hardly find myself at ease [*yo, aquí, poco me hallo*], but what is one to do? One has to stay. (structured interview transcription 2000)

When I asked Don Julio to elaborate why he could not find himself at ease, he replied:

> The thing is that it's not enough. We are between two vacant lots, and they don't want to sell us another one. They say that no, that you can't have two lots, and I say, what they want is the money [an allusion to bribery]. These lots, they're really small. Look, from that post to that post, small. The ones in Marcelino, those are really big. In that *colonia* [suburb] they have progressed a lot. Look, over there they have electric power, and even potable water—everything. (structured interview transcription 2000

Other residents attributed their inability to experience *hallarse* to the conditions of social insecurity that had come to plague Limón over the course of its construction. In the following interview excerpt, another resident, Doña Xiomara Carranza, conveyed this feeling as she explained her life in Limón in 2000. Doña Xiomara had not benefited from the home reconstruction programs and continued to live in a temporary shelter despite the presence of three hundred vacant houses in the site. She shared:

> We don't feel tranquil because we live with fear. We live with fear in the night, and sometimes you start to think, at what time will they

[*mareros*] come to harm you in the night? And at times, you don't sleep, because these micros are not very safe; with one blow you can break it. And one thinks of the children, [hoping] that the delinquents won't hurt them. Because now they don't respect, they don't even respect the children. (structured interview transcription 2000)

In this chapter, I use *hallarse* as an ethnographic point of engagement for exploring the ways disaster survivors invoke sensory experience when they assess the success and relevance of disaster reconstruction projects and how sensory experience is figured (or not) in the ways NGOs and local governments conceive disaster reconstruction projects. In doing so, I also show how disasters and post-catastrophe reconstruction, with their profound impact on built, natural, and social environments, create circumstances where people reflect on their sensory experience of space and place and where they attempt to bring into the discourse that which is supposed to be "unqualified intensity" (Massumi 2000).

For practitioners, NGO project managers, and governmental officials, these reflections are important because they show how affect is the dimension of human experience where recovery projects either become relevant and meaningful for disaster-affected populations or perpetuate the undesired socio-material effects of disasters. While Limón de la Cerca may be a single case, the practices of donor organizations featured in this chapter are not unique or idiosyncratic. They are part of a larger practice of development that operates on very similar premises to those in "Flood Plain Management": Development is a simple process of transferring technologies and governance techniques from one site to the next, this process has predictable and uniform outcomes across localities, and politics and ideas about otherness (e.g., class, gender, national identity, race, ethnicity) do not play a role in it (Escobar 1997; Ferguson 1994; Ferguson and Lohman 1994; Mitchell 2002).

The ethnographic examples featured in this chapter show that development practices are never applied in a sociopolitical vacuum. They are always interpreted and reconfigured by people who are tied to location-specific histories of class, nation, gender, and race formation. The task at hand is to understand how in disaster-affected localities the development activities that are part of a broader cultural horizon of modernization and governmentality become entangled with postcolonial *orders*—a term I use to convey the idea that colonial sociopolitical arrangements did not vanish with independence movements but continued to unfold in postindependence societies (Mbembe 2001; Thurner 2003)—and what implications these entanglements have for the disaster survivors' sensory experience of recovery.

Affect, Space, and Embodiment

In the preceding conversations, Cholutecans spoke about *hallarse* as something they sensed; they either felt *hallados* or they did not. But what generated such feelings among Limón residents? These disaster survivors tell us that the experience of feeling *hallados* is evoked by a relationship between their bodies and those of others, as in the proximity of trusted neighbors and relatives. They also mention the relationship between their bodies and particular kinds of spaces with unique distributions, dimensions, and aesthetics, such as houses with internal partitions, tile roofs that dissipate heat, and larger land parcels that allow for expansions to accommodate extended families. The words of these Cholutecans bring up Gregory Seigworth and Melissa Gregg's (2010) observation that affect is fundamentally something that manifests in the relationships among people and between people and things; therefore, it is not inherent in any one body. These words also resonate with Setha Low's (2011) theorization of the person as a spatiotemporal formation that realizes space in the act of bodily movement, meaning that people make spaces and spaces, in turn, make people.

It is also noteworthy that the experience of *hallarse* is not a universal human phenomenon. I, for example, did not use a word like *hallarse* to reflect on my sensory experience of Choluteca, nor did I share the displaced Cholutecans' affinity for the kinds of spaces and social relations they reminisced about in Limón de la Cerca. Consequently, in this chapter I make the case that *hallarse* was felt by sensing bodies that were shaped over the course of life experiences in the spaces, social relations, and temporal structures that were unique to the affective ecology of pre-Mitch Choluteca and that this ecology was connected to broader transnational interconnections that had increasingly involved the United States during the preceding century.

But this is only half the story. The other half concerns these questions: If the feeling of *hallarse* is the criterion through which disaster survivors in southern Honduras assessed the success and relevance of reconstruction projects, and if the feeling of *hallarse* is contingent on the (re)creation of spatial conditions and social relations that evoke such a sentiment, what conditions were created in Limón de la Cerca that did not evoke a sense of *hallarse* among disaster survivors? Furthermore, who created them and why? Finally, how do answers to such questions stand to inform disaster recovery practice in localities beyond southern Honduras?

Over the course of my ethnographic research, I spoke with local and international NGO project managers, municipality officials, and displaced and non-displaced Cholutecans; spent time in the two resettlement sites, observing community activities and projects; and collected project evaluation documents and reports with the intention of eliciting a microhistory of the reconstruction process. During these conversations I asked my interlocutors whether they perceived any differences between Limón and Marcelino and, if they did, to explain why these differences existed.

The NGO project managers I spoke with explained the difference between the two communities as the manifestation of disaster survivor values and social organizing practices before the storm. The people of Limón, I was told, lived in Choluteca's marginal neighborhoods and were themselves marginal people, with a propensity toward delinquency and a lack of social cohesion. The people of Marcelino, in contrast, had a rural provenance and an established history of community organizing; and as such, their resettlement site was doing so well.

These explanations fit very comfortably within some theorizations of resilience in the community psychology literature. In these definitions, a community's *social capital*—a term used to denote the number and strength of social interconnections among members—and other pre-disaster "capacities" are thought to play principal roles in its ability to recover from a disaster (Foster-Fishman, Pierre, and Van Egeren 2009; Kusel 1996; Norris et al. 2008; Sherrieb, Norris, and Galea 2010). The case of Marcelino and Limón, however, presented a challenge to such models. A survey I conducted of 110 randomly selected households in Limón and 50 randomly selected households in Marcelino (with the assistance of Honduran research assistants Rosa Palencia and Carlos Giacoletti) indicated that the people who lived in these two sites predominantly originated from twenty-two different *barrios de clase obrera* (working-class neighborhoods), to use the description employed by local public health officials. Immediately after the catastrophe, these displaced Cholutecans attempted to act as a single group and eventually split into two distinct communities in 1999 as an effect of political struggles between the municipal government and a nascent disaster survivor grassroots leadership.

Explanations for the community's differences that relied on the idea that Limón and Marcelino existed as distinct groups prior to the disaster, with discernible levels of social capital and community capacity, did not hold. Marcelino and Limón, in fact, emerged as communities over the course (and as a result) of the reconstruction process. Moreover, as the following ethnographic narrative shows, the social and material conditions that took shape in Limón and Marcelino could not be solely attributed to any inherent capacities of community residents; instead, they came about as an outcome of the politically and epistemically mediated relationships between national and international aid organizations, government agencies, and the displaced. What follows is the ethnographic narrative I composed through the hundreds of formal and casual conversations about the resettlement that I had with the aforementioned people, followed by a reflection on the kinds of spaces and *socialities*—a term I use to denote the qualities of relationships among people—that emerged through the reconstruction process and on the disaster survivors' affective experience of them.

*Spaces, Socialities, and Affect in
the Aftermath of Disaster*

In the beginning of this chapter, Don Julio Rodriguez assessed his life in the resettlement site of Limón de la Cerca by saying "yo, aquí, poco me hallo" in direct reference to the dimensions of the land parcel for his house and the constraints that Cholutecan municipality reconstruction policies placed on his ability to acquire adjacent properties to expand his diminutive house. To understand Don Julio's sentiment of not being at ease, it is necessary to have an appreciation of the history of household and neighborhood spaces in Choluteca prior to Mitch and the notions of modernization and capitalism reintroduced in the catastrophe's aftermath by USAID and other assisting NGOS. I write "reintroduced" because the ideas, techniques, and practices of development and disaster recovery I am concerned with were not completely novel in the region, but as in many other cases (Gunewardena and Schuller 2008; Klein 2007), the disaster enabled national political leaders and aid agencies to declare a state of emergency that permitted yet another wave of their extensive and forceful deployment.

The area that is today's city of Choluteca was occupied by an indigenous settlement in pre-Columbian times. In the 1520s Spanish colonial administrators declared the site a mining concession and ordered the creation of a *reducción*, a term used in colonial Latin America to refer to forced relocations of indigenous populations who either lived in scattered village settlement patterns or were actively distancing themselves from areas of Spanish control to avoid the excessive tribute, forced labor, and violence imposed by Europeans. The idea behind the *reducciones* was to make the labor and tribute of indigenous populations readily accessible to the Spaniards who were given rights over them as part of the *encomienda* system. In this colonial practice, appointed *encomenderos* (holders of the *encomiendas*) were "entrusted" by the Crown with the Christianization of local populations. Indigenous populations, in turn, paid tribute to the *encomenderos* as repayment for the "service" of their Christianization.

Colonial settlers built Choluteca alongside the similarly named river, which flows from the highlands of the Central American isthmus and drains into the Gulf of Fonseca. The town was constructed following the Spanish grid style with a central plaza flanked by a church and an *ayuntamiento* (the local government offices.) This plaza, which symbolized and materialized the social order of colonial Central America through its religious and colonial administrative constructions, sits on higher ground south of the river. Today it is surrounded by well-to-do residential areas, while more modest *clase obrera* households are located on the Choluteca River floodplain directly adjacent to the river.

The *gente de clase obrera* (working-class people) who resettled to Limón and Marcelino devoted themselves primarily to work as ambulatory food vendors, maids, housekeepers, laundry washers, construction workers, truck and taxi drivers, agricultural wage laborers, security guards, and operators of small, home-based convenience stores called *pulquerías*. Some residents of *clase obrera* neighborhoods were part of families that had resided in Choluteca for multiple generations, while some recent arrivals were dispossessed subsistence farmers who had migrated to the city from the southern region's rural areas.

While the Honduran National Census officially designated Choluteca as a mid-size city, the use of the term *city* requires some reflection on the relationship between the language that demographers use to speak about spaces and the variation that exists among the built environments that are designated through such terminology. In Choluteca's pre-Mitch urbanism, people had adapted a variety of practices such as animal husbandry and horticulture to the densely settled area, and these practices had implications for how space was structured in subaltern neighborhoods. Houses featured outdoor spaces where families planted house gardens and grew tomatoes, chilies, fruit trees, and cooking and healing herbs, and where the occasional pig and chickens roamed.

These Cholutecans also built houses in a piecemeal fashion, adding rooms and extensions one at a time as children were born, grew into adulthood, married or paired off, and had children of their own. The homes were built with a combination of materials procured from the local environment (e.g., sand and clay from the Choluteca River fashioned into construction blocks and roof tiles) and mass-produced commodities such as electrical outlets. A case in point was the house of Doña Juana Rodriguez, the spouse of a Choluteca policeman whose home was located in the flood-damaged neighborhood Barrio Las Cruces. Although many of her neighbors were forced to relocate as a result of the destruction caused by Mitch, the residence she lived in only sustained minor damage, and her family was able to remain in the city. Doña Juana, a woman in her early thirties, lived in an extended household with her in-laws. Her husband, father-in-law, other male relatives, and hired workers had built her house one room at time as the family grew.

While some Cholutecans from *barrios de clase obrera* lived in one house over several generations, others outgrew their houses and constructed and moved into larger ones as their families required, and they

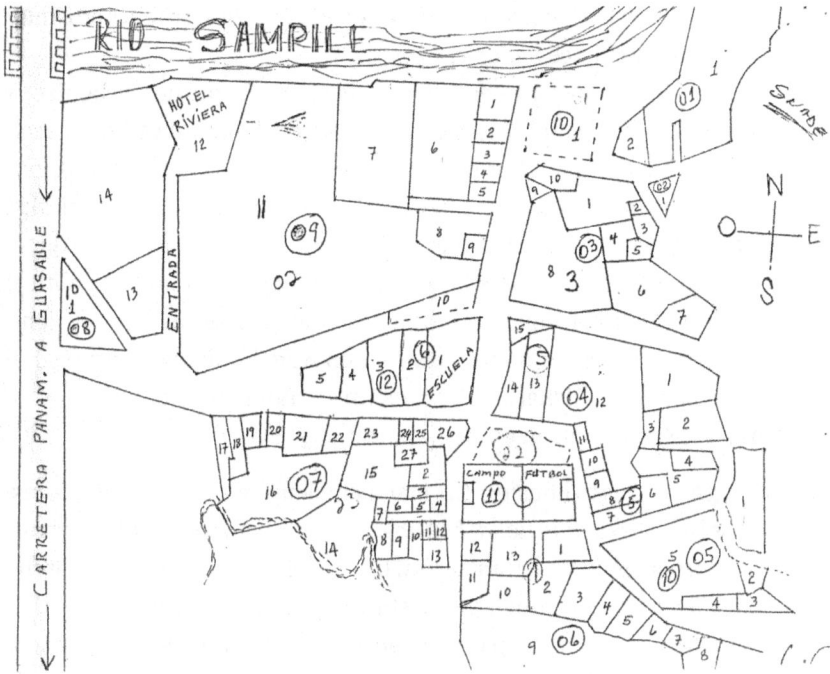

FIG. 8. Spatial distribution of a Choluteca neighborhood before Hurricane Mitch. Courtesy of the Honduran Institute of History and Anthropology.

turned the older structures into rental properties. Such was the case of Don Rodrigo Portillo, a Cholutecan man who, along with his wife and three adolescent children, was displaced during Mitch. Prior to the hurricane, Don Portillo had built three homes over the course two decades, and these structures, which were all destroyed by the flood, had represented his life's labor and social standing.

Because Cholutecans' homes changed over time with the organic growth of the families who lived in them, so did the layouts of their neighborhoods. As figure 8 shows, the varied size and shape of land parcels partly represented the growth of families. The spaces of these *barrios de clase obrera* were, to an extent, material biographies of the families who lived in them. Furthermore, as Pierre Bourdieu (1977) and Setha Low (2011) have observed, such spaces, along with the social relations that took place in them, became embodied over the course of people's life experiences within them. These spaces played a role in shaping how people experienced their sense of comfort and aesthetics and their ability to experience *hallarse*.

The U.S. Agency for International Development did not recognize or appreciate the spatial-biographic and affective nature of Cholutecan neighborhoods in the aftermath of Hurricane Mitch. Because the agency operated from an epistemic position of U.S. market-based capitalist notions of homes as mass-produced commodities, USAID program directors considered Honduras lacking for not having a familiar political economy. Hence, from USAID's vantage point, Honduras was deficient, and local practices of home construction were by and large ignored in recovery efforts. Richard Wilson, USAID's national chief of housing programs, explained in an ethnographic interview:

> With regards to housing, the major challenge was that, in Honduras, there is *no real functional housing market*. That was one of the greatest problems in structural terms. Instead of channeling money for the reconstruction through the existing mechanisms, we had to create a market. We created a parallel housing market, which will disappear when our funding ends. (structured interview 2001, emphasis added)

Because home construction practices in Honduras were not akin to those of the United States, where many homes are built by third-party contractors and sold on a retail market, and because USAID personnel ethnocentrically figured Honduras as deficient, the agency defined the challenges of disaster recovery accordingly. Its primary concern, then, was to create a familiar capitalist economy of home construction and sales rather than assist in creating those spaces that would be conducive to displaced Cholutecans' experiencing *hallarse* and recovery. The economics of mass home construction became both the primary concern of USAID in post-Mitch Honduras and the primary mechanism for evaluating housing reconstruction projects. USAID project evaluators deemed housing projects like Limón de la Cerca successful if project budgets were spent on time and on their designated purposes (or, at least, if the Excel spreadsheets of project reports claimed this was so), but they did not consider whether the built environments created with these funds made sense to disaster survivors.

I became familiar with these practices of disaster recovery management when I made an agreement with a USAID public relations officer to share a conference paper I had written about Limón. The paper—an early draft of this chapter—focused on issues concerning the inadequacy of housing design in southern Honduras's resettlement projects. During a Christmas gathering at the U.S. ambassador's house in Tegucigalpa in 2000, the USAID public relations officer commented that a team of evaluators had just returned from Choluteca, where it had conducted an assessment of Limón. I was eager to see what knowledge the evaluators had about the progress of reconstruction projects and what marks the site had been given. We agreed to a document exchange: I would give her my conference paper, and she would share the evaluation materials with me. After sending my paper via email, I received the USAID evaluation in return. The evaluation comprised a very elaborate spreadsheet that tracked budget expenditures in Limón but no accompanying text. According to the budget, the creation of a secondary housing market in southern Honduras was coming along splendidly, and reconstruction was on track.

The USAID public relations officer also agreed to disseminate my conference paper among agency personnel, and within a week I received an email from the national chief of housing, who doubted my claims about faulty home construction. In response, I walked to the houses of a few displaced Cholutecans who were injured or killed when the donated homes had structural failures, and I collected death and medical records proving the residents' injuries. I scanned the documents and sent them to AID, requesting that the families of the dead and injured be compensated for their losses. In response, I was told USAID would send a team to inspect the structures in Limón and reinforce them if necessary.

The contrast between the budget and the ethnographic narrative of a conference paper brings up the politics of knowledge in the governmentality of disaster reconstruction. As Richard Bauman and Charles Briggs (2003) have noted, a key development in the formulation of modern ways of knowing was the creation of a scientific language that was allegedly detached from its social context and the discrediting of other socially embedded ways of speaking and writing. Modern forms of knowing necessitated the creation of systems of documentation that were thought to represent "the world as it is," irrespective of cultural prejudices or emotional overtones. The budget represents just such modest witnessing language. USAID project administrators credited numbers, instead of words, as a more rational and objective means of apprehending the world of disaster recovery. This language translates the lived experience of disaster survivors into a sanitized numerical chart that leaves out the messiness of the politics, violences, and affective discomforts that are the life of reconstruction projects.

The power of the budget over the imagination and agency of USAID personnel emerged again and again over the course of my ethnographic experiences in Limón (and continues to haunt me in the academy, as university administrators use it as a mechanism for gauging the "health" of educational institutions). On a bright January morning in 2001, I met another agency evaluation team that had driven to Limón from Tegucigalpa in a late-model sport utility vehicle to conduct yet another assessment of the site. Seeing the team, I approached its members and struck up a conversation about the state of the resettlement community. I brought up the issue of housing design, the spatial qualities of the reconstruction site, and the ways disaster survivors were experiencing the donated homes. Andrea Jones, a USAID project evaluator and consultant replied: "We wanted to take cultural considerations into account, but spending the budget on time became our top priority" (unstructured interview 2001).

USAID program managers, however, argued that some efforts were indeed made to construct Limón de la Cerca in a way that took sociospatial considerations into account. The agency's national chief of housing programs outlined these considerations as part of what the agency called "integrated solutions":

Housing projects were required to be permanent solutions, with an emphasis on what we called integrated solutions. The "integrated" part refers to the integration of the home within its spatial conditions. Homes were meant to be more than just a housing structure. Constructed homes were required to be accompanied by potable water [which was not yet present in Limón over the course of my ethnographic stay], access to public ways, latrines, and sewage [also not yet present], and that the reconstruction area not be in a zone of high risk. (structured interview 2001)

The integrated solutions approach, however, did not guarantee that the spaces, structures, and social relations that manifested in Limón would be familiar and make sense to disaster-displaced Cholutecans. Instead, it was a template whose execution satisfied program requirements, irrespective of the ways its socio-material effects were experienced by the people to whom it was applied.

Politics, Knowledge, and Social Relations in Disaster Reconstruction

The construction of Limón de la Cerca's built environment cannot be solely reduced to the capitalist and modernist disaster management tendencies of USAID's disaster response. Unlike the tidy image of development as policy and technology transfer conveyed by Cuny's (1983) "Flood Plain management," the reconstruction encounter is a rather messy entanglement of locality specific, historically contingent, and transnationally engendered political culture, development practice, and material agency. It is to this messy entanglement that I now turn the attention of this ethnographic narrative.

The social disruption caused by Hurricane Mitch and the limitations of local and national governments in Honduras—as in many other disaster-affected sites—created a moment when grassroots organizations emerged out of preexisting civil society groups and set out to handle the unaddressed logistical and social challenges of reconstruction. When Mitch struck, displaced Cholutecans relied on their kinship and neighborhood relations to respond to the catastrophe. Immediately after the storm, future residents of Marcelino and Limón followed similar patterns in search of shelter. Most sought refuge in the city's schools, but a substantial number also found haven in the homes of friends and relatives, and a smaller fraction settled in neighborhood churches. Families who sought temporary housing in schools remained there for three months, but eventually administrators pressured them to vacate the premises for the coming school year. Given the municipality's sluggish response, a small group of disaster survivors who had previously participated in neighborhood civil society organizations called *comités de desarrollo* (development committees) decided to take matters into their own hands. With the assistance of a Spanish expatriate Marist religious brother, Manuel Fernandez, these grassroots organizers began an independent search for a place to resettle.

The group identified a property that would eventually become Limón de la Cerca approximately seven kilometers to the northeast of the city along the Pan-American Highway. Various groups had attempted to develop the land for agricultural purposes, from cattle ranching to rice farming, but none of these initiatives had proven successful. At the time, Banco de Occidente, a major national bank, possessed the land's title. Grassroots leaders found the 840-acre property attractive because of its low price. Residents could then purchase 4,300-square-foot land parcels, which disaster survivors felt was a necessary size to accommodate their daily habits. Land parcels had to be large enough to allow future home expansions because, as noted previously, families as they grew customarily added rooms to their houses. They also required additional space surrounding their new homes to continue the common practices they had pursued in their flooded neighborhoods of Choluteca: small animal husbandry, vegetable gardening, and the planting of fruit trees.

Despite their initiative, grassroots leaders were limited in their capacity to act as legal representatives of displaced Cholutecans and found themselves, once again, on the municipality's timetable. Faced with continued procrastination on the part of local government, grassroots organizers urged other displaced Cholutecans to follow them on a partial land invasion of the site in January 1999. Not wanting to create unnecessary frictions with the owning bank, the disaster survivors built a small shantytown along the sides of the Pan-American Highway, hoping their precarious condition would call attention to the municipality's continued delays. Don Carlos Carrales, one of the grassroots leaders and a member of Limón's resident-staffed *comité de desarrollo* recollected this moment:

> Each shelter had people from different neighborhoods, and each neighborhood had its own leaders from the time before the disaster. These leaders were some of the first people to survey the area where Nueva Choluteca [another name given to Limón de la Cerca] was constructed, and they began the land invasion. Each leader brought their own people from the shelters. They came little by little, and

then, when the people in the shelters saw what was going on, they started to come. Some of the leaders that are in charge now were neighborhood leaders before the hurricane; some weren't. Some became leaders out of necessity; others became leaders to see what they could gain. (structured interview 2000)

Living on the shoulder of the Pan-American Highway posed a new risk for Choluteca's displaced. Within a month of the limited land invasion, a young girl was hit and killed by a car. Her death acted as a catalyst on the mounting frustrations of displaced Cholutecans, who placed responsibility for their undesirable living conditions on the municipality's slow response. After the child's death, disaster survivor leaders organized a protest with the intention of pressuring local government into action. The protest took place in the city's central area, with protesters blocking the entrance to Choluteca's most prominent landmark, the Choluteca River Bridge. Don Carlos Carrales explained the motives behind the protest:

We were housed in various shelters, and we didn't know where we were going to end up living, so a group of people decided to take those lands. The Catholic Church contributed tents so people could live there. In the end, we took the bridge so the mayor would finally tell us where we could go resettle. Because of the mayor, we lost a lot of aid. The German Red Cross went to Marcovia [another town approximately five miles west of Choluteca] because the mayor had not resettled us. At last, a land commission was formed, but to this date, we don't know how much was actually paid for the land. The mayor handled everything behind closed doors. There is a document, a bill of sale, but we have not seen it. Maybe you can talk to the lawyer who handled the sale so you can see it. These lots, they were supposed to be affordable. (structured interview 2000)

The protest triggered a reactionary response on the part of the mayor, who summoned local police to dismiss it. The attempt to disperse the protesters turned violent. Police officers beat several disaster survivors and arrested protest organizers. Following these events, the mayor decided to take a more proactive role in the acquisition of the resettlement site. He appointed a land committee of municipality workers and three disaster survivor organizers who resided in Limón and who (according to their fellow displaced Cholutecans) were co-opted with the offer of free land and houses in the reconstruction site.

The municipality-appointed land committee made a number of policy decisions that had profound implications for the construction of Limón de la Cerca. Rather than proceeding with the 4,300-square-foot land parcels that the grassroots leaders preferred, the committee opted to reduce the size to 1,300 square feet with the rationale of assisting a greater number of families. The committee also decided to randomize the land distribution through a raffle, arguing that such a practice would ensure equity and transparency. At the same time, at the mayor's behest, the land committee excluded those proactive grassroots leaders whom the municipality perceived as a political threat.

Over the course of my ethnographic research, government officials defended these actions on the basis that cost-benefit logic (i.e., reducing the size of land parcels and therefore housing more families) and transparency measures (randomized parcel distribution) were unquestionable and self-evidently rational governance practices. Nevertheless, the land commission's resettlement policies became a means for local political leaders such as Choluteca's mayor to exercise political power over those disaster survivors who had strayed from the expected norms of local political culture.

In January 2001 Manuel Fernandez, the Marist religious brother who had assisted the grassroots organizers, recalled this process:

> The mayor raffled the lots at El Chilo [one of Choluteca's schools]. That was a lie and a farce! The leaders were marginalized and did not receive any lands. The mayor bought off a couple of the community leaders by guaranteeing them homes and continued with the raffle. (structured interview 2001)

The literature on the political culture of Honduras tells us that the mayor's actions are an example of a broader governance practice called clientelism, or the use of minimal gifts in patron-client relationships to secure political support (Allison 2006; Euraque 1996; Rosenberg 1988; Sieder 1995). In Honduras, clientelist practices date back to the early postindependence period, when regional patriarchal charismatic leaders called caudillos used gifts as a means of consolidating their political power. This practice received a new life in the twentieth century when U.S. capitalist interests (i.e., fruit companies) relied on clientelism to steer Honduran national economic policy, making clientelism a hallmark institution of governance in the country (Euraque 1996; Sieder 1995). Through the Cold War, cycles of clientelist gift circulation were further entrenched in Honduran political and community life (Sieder 1995), and these practices became one of the primary mechanisms through which local government and some assisting NGOs related to disaster survivors in Limón.

Like the disaster-affected communities, national communities themselves came into being in the midst of global power relations that entail colonialism and Cold War and new world order politics. The U.S.-based evangelical NGO working in Limón at the time, Samaritan's Purse, for example, was directed by Deborah DeMoss Fonseca, a former U.S. Senate Foreign Relations Committee staff member who had served under Jesse Helms and married a prominent Honduran military officer with presidential aspirations (*New York Times* 1994). It is also noteworthy that the founder of Samaritan's Purse is Franklin Graham, son of evangelist Billy Graham, and that the organization became known in Honduras for its right-wing political tendencies, as DeMoss's congressional experience evinces. The assertive actions of Choluteca's displaced threatened to disrupt a deeply rooted transnational system of political culture, prompting the reactionary actions of local government.

The municipality's land committee practices had profound implications for the construction of Limón. The random distribution of land parcels fractured important social relations between neighbors and relatives who had lived in proximity to each other before the storm. These relationships were central to the experience of *hallarse* among disaster survivors. Ethnographic observations in hurricane-affected *barrios de clase obrera* in Choluteca showed me that longtime neighbors once relied on each other for assistance with child care and household security. In Limón, in contrast, residents reported not feeling enough *confianza* (trust) with their current neighbors, and the conditions of relative anonymity contributed to working mothers' difficulties in finding child care and to residents' lack of willingness to assist one another when victimized by gang members.

The municipality's intervention not only spatially divided neighbors and kin groups but also politically divided disaster survivor leaders into two camps—those who sided with the ousted organizers and those who acquiesced to the policies of the municipality land committee. With the assistance of Manuel Fernandez, the ousted leaders and approximately six hundred displaced families founded a second resettlement site only a few hundred feet west down the Pan-American Highway and named their nascent community Marcelino Champagnat, the founder of the Marist religious order.

The actions of the municipality had a polarizing effect on Limón and Marcelino, making a docile social body of the former and spurring the formation of a robust and assertive network of grassroots organizers in the latter. This difference in community organization had significant implications for executing housing reconstruction and infrastructural projects at both sites. In Marcelino, assertive grassroots leaders became renowned for their acts of resistance to those aid projects they found socially and environmentally unsuitable to the particularities of southern Honduras. These residents proudly retold the story of how community leaders rejected a proposal by the NGO CARE to build housing structures similar to those of Limón. One community organizer recalled these events:

> We told them we did not want their matchbox houses and that we would rather stay living in the *conos* [tents], and two weeks later, they came back with plans for larger houses and with four rooms! (structured interview 2001)

This act of resistance encouraged CARE project managers to reconfigure their project budget, and they reduced the cost of nonessential personnel and allowed for the construction of houses with larger floor plans, internal partitions, and reinforcing columns. The CARE project manager in charge of housing construction programs in Marcelino narrated these events thusly:

> I participated in those meetings, especially the meeting where the disaster survivors asked that the houses be made bigger. You know how those meetings are; sometimes people get excited. The people said that they did not like the design. They said they were not used to sleeping

[and living] in one room and that the small size of the houses was not suitable. With CARE, we eliminated the manager and the architect, and we hired five foremen and one senior construction worker per house. The construction assistants were the disaster survivors, and that way the houses came out more affordable. (structured interview 2001)

In Limón, in contrast, architects working for Samaritan's Purse routinely mobilized discourses of cost-benefit analysis to reject disaster survivor requests for alternative home designs and construction techniques that would allow disaster survivors to experience a sense of finding themselves at ease. My household survey, for example, indicated that 27 percent of adult men who lived in the resettlement site earned a living as construction workers and had knowledge of construction techniques. Over the course of ethnographic interviews, these men pointed out critical structural weaknesses in their newly constructed homes. Don Ramón Rosa, an experienced construction worker and resident of Limón, explained these shortcomings during a conversation:

A well-constructed house has four columns, and the horizontal beams go on top of them, but not [in] these houses. They only have a piece of rebar that goes through the cinder blocks on the corners, and I told the construction foreman of Samaritan's Purse, but he didn't want to listen to me. They said it couldn't be done because of cost-benefit. (semi-structured interview 2000)

The structural weaknesses of Samaritan's Purse–constructed houses would prove deadly. The resettlement site was located on a semiarid plain, where small wind funnels and strong thunderstorms routinely ripped poorly attached roofs from houses and made unsupported walls collapse on their residents. Despite the claims of USAID's integrated solutions approach, the combination of political culture, cost-benefit analysis, cheap construction, and environmental particularities ensured the disaster continued long after Mitch dissipated.

Choluteca's Modern Urbanism

Limón's construction followed a master plan drafted by one of Honduras's principal engineering firms, Nacional de Ingenieros. The plan called for arranging housing structures and land parcels in neat rows along straight streets that collectively formed a diamond shape when viewed from above (see fig. 2 in "Introduction"). The imagination that made this master plan possible has a history that ties post-Mitch Honduras to distant times, people, and places. Limón's architectural blueprint followed conventions of modern urbanism, which emerged in western Europe during the nineteenth century. A key idea behind modern urbanism is the notion that geometrically regimented space can produce "normatively" behaving people—that is, disciplined laborers, soldiers, and reformed convicts who can follow strict temporal schedules and act in unison under a central command (Holston 1989; Rabinow 1995). The community design drafted by Nacional de Ingenieros staff implicitly proposed that the modernist regimentation of space in Limón would be conducive to the production of social well-being and recovery, but the spaces this design engendered were incapable of evoking a sense of *hallarse* among displaced Cholutecans.

While the people of Limón could not experience the social and material conditions taking shape in Limón as recovery, the Cholutecan municipality mobilized images of the resettlement site (with its straight rows of houses and rigidly regimented space) as indicators that successful participatory development and reconstruction had been achieved (fig. 9). The municipality's 1998–99 government report, for example, features a two-page layout that reads "Participation Towards Reconstruction" and displays images of housing structures as evidence that recovery was proceeding effectively. Missing in these images were the disaster survivors, the street-gang violence, and the abandoned and storm-damaged homes that pervaded my ethnographic experience.

Limón de la Cerca in Global Context

During a recent sabbatical research stay in Mexico, I was invited to conduct a presentation about my research in southern Honduras at the National Polytechnic Institute's School of Architecture, where architects and anthropologists have collaborated in planning and evaluating a number of post-disaster resettlement communities. I titled the talk "Digamos que Yo, Aquí, No Me Hallo" (Let's say I don't feel comfortable here). Following the presentation, one of the architects who has worked

Participación para la
Reconstrucción

En 1998 todo era esperanzador en Choluteca, la economía crecía y la capital de la agro industria visualizaba un futuro promisorio. Pero todo este potencial económico, fue reducido por el fenómeno nacional MITCH. La ciudad fue afectada en un 18 por ciento, equivalente a nueve barrios totalmente destruidos y ocho en forma parcial.

Informes estadísticos reportaron tres mil viviendas afectadas, 167 muertos y 500 millones de lempiras en pérdidas en activos de viviendas, a esto se suma el cementerio que quedó totalmente destruido.

La industria camaronera fue afectada en un 100 por ciento, la melonera en un 40, la azucarera en un 70, las vías de comunicación en un 60. También fue afectada los granos básicos y ganadería. La inundación provocó una evacuación de 39,200 personas, los cuales se ubicaron en 202 albergues provisionales: escuelas e iglesias y municipalidad.

La Corporación de Choluteca se convirtió en un ente facilitador através de Compra y gestión de terrenos donde se construyeron las viviendas a los ciudadanos reubicados.

En la ciudad nueva se compraron 114 manzanas de terreno por valor 3,573,000.00 millones de lempiras, don-

En la reubicación Colonias Unidas, se asentaron 1500 familias, donde ya se han construido 120 casas por parte de la iglesia evangélica.

Vista panorámica de Ciudad Nueva donde han sido beneficiadas más de mil familias.

P.18 INFORME DE GOBIERNO 98/99

FIG. 9. Choluteca annual government report, 1998–99. Courtesy of the Honduran Institute of History and Anthropology.

on a number of resettlement projects in Yucatán and the Gulf Coast of Mexico commented on the title: "I found your use of *no me hallo* very interesting; it's something we keep hearing in the resettlement communities" (fieldnotes 2013).

The case of Limón de la Cerca is relevant to the literature on disaster reconstruction not only because it highlights the importance of recognizing the affective dimensions of community resettlement and development but also because Limón is by no means an idiosyncratic or isolated case. In the time following my fieldwork, I learned of more and more cases of post-disaster resettlement that are strikingly similar to that of Limón. These case studies have repeatedly revealed instances when disaster-affected populations are channeled into modernist urban designs (Audefroy and Cabrera 2014; Briones 2010; Ganapati and Ganapati 2009; Macías 2009), where their social relations are fractured by the random distribution of land parcels and where their requests for alternative practices of community (re)construction are denied by NGO or governmental agency workers who rigidly use budgets as a means of program management and evaluation.

While the case of Limón shows us the implications of modern urbanism and of what Frederic Jameson (1991) would call "late-capitalist" disaster reconstruction management, the case of Marcelino also shows us an alternative where NGO program managers and disaster survivors engage in a dialogical process that is conducive to the creation of communities that are relevant in affective, material, and social terms to disaster survivors. In the case of Marcelino, we see CARE project managers who practice what I call epistemic flexibility. NGO project managers in Limón routinely invoked and fetishized the budget as an unmodifiable mechanism of reconstruction program management and assessment. Such invocations legitimized why housing reconstruction programs could not be tailored to the spatial and social arrangements that the displaced Cholutecans were accustomed to. The budget operated as a key node in a collection of power relations between disaster survivors and aid agency workers, a node that was buttressed by ideas about modernization as an unquestionably desirable outcome of disaster reconstruction and that viewed expert knowledge as a factual and unchallengeable discourse.

In Marcelino, in contrast, CARE program managers recognized the importance of the budget as an instrument of disaster recovery, but they also kept in mind the budget is the product of human action. This approach allowed them to exercise a different kind of agency, modifying its structure and content to suit disaster survivors' self-defined needs. The act of being epistemically flexible—that is, the willingness to accommodate disaster survivors and to recognize that the budget is the product of human action—had affective implications for displaced Cholutecans as the reconstruction effort allowed for the creation of socio-spatial arrangements they found relevant, pleasant, and understandable.

3. Feelings of Inequity

GENDER AND THE POSTCOLONIAL MODERNITY OF DISASTER RECONSTRUCTION

Before Hurricane Mitch, Doña Concepción Rodriguez lived in Barrio Buenos Aires, one of the riverfront neighborhoods of Choluteca. She shared a home with her husband, Don Ricardo, and their five children and helped support her family by making and selling tortillas in the city's central market. When they needed assistance with child care while she was working outside the home, for many years the Rodriguezes relied on trusted neighbors who lived in close proximity. They had to feel *confianza* (trust) with someone before entrusting their children to that person's supervision. In southern Honduras, *confianza* is a sentiment that emerges over time, but time is not the only element involved in the sensation. For people to develop *confianza*, Cholutecans like Concepción and Ricardo also have to sense a kindredness with the cultural embodiment, the way of being, of their neighbors.

Two and a half years after the catastrophe, I sat down with Doña Concepción at her home in Limón de la Cerca and talked about her life in the resettlement site. During our conversation I asked about the ways she and her husband made a living in Mitch's aftermath, and she replied, "I work here in the house, watching the children. I don't work" (structured interview transcription 2000). When I asked why she no longer sold tortillas at the market, Doña Concepción explained, "Now I don't sell because I am afraid of leaving the children, leaving the children alone, because here there are many delinquents." In addition to her fear, which was generated by the conditions of social instability in Limón, Doña Concepción had also abandoned her work at the market because she had yet to develop a sense of *confianza* with her new neighbors. The municipality's random distribution of land parcels had fractured important affect-laden relationships among longtime neighbors and relatives, and although Doña Concepción found some of her new neighbors to be agreeable, she had not yet developed the trust necessary to leave her children in their care. I asked her how she felt about this new state of affairs (not working outside the home), and she replied:

> I miss it, yes, I miss it, because it's not the same thing as one working, helping your partner. Now it's only him; he's the sole head of household [*cabecilla*]. He now maintains us, and before I helped him, but now it's only he who looks after us. No, you feel more, well, different. It's not the same. (structured interview transcription 2000)

Unfortunately, Concepción's experience of disaster reconstruction is not unusual. With regard to gender and disasters, social scientists such as Susana Hoffman (1999) and Sarah Bradshaw (2001) have documented how, in the aftermath of catastrophes, women often not only experience a devaluation of their labor, especially that done in the home—as demonstrated by Concepción's declaration "I don't work"—but also find their access to political and social spaces outside the home diminished. Research has also found that women often confront social circumstances that place them at greater risk of bodily harm, increased economic dependency, limited social mobility, unfavorable labor circumstances, political marginalization, and disruption of their lifeways than men do in disaster contexts (Anderson 1994; Bradshaw 2001; Cupples 2007; Enarson 1998, 2000; Ensor 2009; Greet 1994; Hastrup 2011; Hoffman 1999; Maybin 1994; Stehlik, Lawrence, and Gray 2000; Walker 1994). Collectively, the existing literature on gender and disasters demonstrates that gender is a fundamental dimension of disaster vulnerability, with women consistently experiencing enhanced levels of risk when compared to their male counterparts; that there are a number of important connections between development practices and vulnerability; and that disaster mitigation policy must prioritize a focus on gender. Despite these insights, recent assessments of governmental and nongovernmental organization disaster recovery practices indicate that there is yet much to accomplish in the mainstreaming of gender in disaster relief (Ensor 2009).

When thinking about the experiences of women in post-disaster contexts, some social scientists have interpreted the enhanced inequities between men and women as a resurgence of traditional gender roles

(Anderson 1994; Ensor 2009; Hoffman 1999). The word *traditional* in this instance is used to refer to a condition where women's spatial and social mobility is curtailed in comparison to that of men and where men establish a monopoly on household and community leadership. Concepción's appraisal of the shift in power relations in her home—"now he is the only head"—could be said to be a classic example of this resurgence of "tradition."

The use of the term *traditional* invokes a spatiotemporal metaphor in which human experience and relationships are conceptualized as taking place within a linear continuum, where tradition is at one end and modernity at the other, and the transition of people from a state of tradition to modernity is considered "development." In the case of disasters, development is sometimes also seen as synonymous with modernity (as a state rather than a process)—that is, the condition in which industrialized societies have successfully harnessed "natural hazards" and disasters are no longer a threat to human life and property (Cuny 1983; Ensor 2009).

In this chapter, I use excerpts from ethnographic interviews with disaster-displaced Hondurans to explore how conditions of enhanced gender inequity and gender change were not so much a reinstatement of traditional gender roles but were, instead, an outcome of reconstruction practices on the part of assisting NGOs and governmental agencies that articulated implicit assumptions about development and modernization. Rather than being a reinstatement of traditional gender roles, I argue that the curtailed socio-spatial mobility of women in Limón de la Cerca was the effect of a particular kind of postcolonial modernity instituted over the course of reconstruction programs in southern Honduras. Furthermore, I also explore how Choluteca's disaster survivors spoke in affective and emotional terms about their experience of the relationships among people and between people and things manifesting in Limón. Conditions of gender change and inequity, these disaster survivors suggested, were things they felt.

Gender, Development, and Modernity

The literature on gender and disasters calls attention to an intimate link between development and vulnerability (Anderson 1994; Ensor 2009;

Greet 1994), with vulnerability being the idea that human practices can enhance the destructive capacities of geophysical phenomena and place certain socially differentiated sectors of an affected community (e.g., gendered, racialized, or ethnicized groups) at greater risk of bodily harm and sociopolitical marginalization during and after catastrophes. While vulnerability is easily defined, outlining the meanings and lived experiences behind the concept of development is a more complicated matter.

The history of the concept of social development in western European thought can be traced to medieval Judeo-Christian beliefs about the second coming of Christ and the envisioning of history as a linear progression toward this imagined moment of salvation (Fabian 1983). Judeo-Christian ideas about time and history (as a singular and linear process leading toward a known outcome) became the cultural framework through which nineteenth-century social theorists such as Karl Marx, Lewis Henry Morgan, and Charles Darwin thought about cultural difference, time, social change, and gender (Castañeda 2002; Chakrabarty 2000; Foucault 1970). At the heart of nineteenth-century perspectives on development were the ideas that all human societies were simultaneously partaking in a singular historical trajectory toward a common endpoint and that differences between societies could be explained in terms of their progression along stages in this linear path.

Time and space were intimately conjoined in nineteenth-century models of social evolution. To advance in time was to move to a different space, and to transform a space was to advance in time. In an ethnocentric move, however, social theorists of the period usually picked their own national societies and gender norms as examples of the most advanced stages of human history (i.e., England, "Western culture") and incorrectly viewed the societies of colonial dominions, which were culturally different, as being mired in earlier developmental stages (Castañeda 2002).

Ideas about development and developmentalism (the gaze and mindset where human diversity is reorganized along a developmental telos) must also be understood in connection to another set of closely related concepts—modernity and modernization. As I reviewed in chapter 1, *modernity* is first and foremost a claim people make of being able both to objectively see the world through eyes that are free of cultural bias and to describe the world and its phenomena in a language that is allegedly not rooted in any one cultural context (Bauman and Briggs 2003; Keane 2007; Latour 1993b). What is interesting about the concept of modernity is that it also allowed its claimants to think about difference in temporal terms. To think like a modernist and adopt the knowledge and technologies that modern epistemology made possible meant advancing from a temporal space of tradition to an emancipatory era of modernity. With regard to gender, "traditional" gender roles came to be envisioned as a way in which people arbitrarily assigned values and possibilities to human bodies, whereas in a state of modernity, bodies ostensibly became liberated from the "prison of gender" (McClaurin 1996).

But modernist ways of thinking about bodies and gender may not have been emancipatory at all. Modern epistemology did not so much lead to liberation from gender as it led to a redefinition of gender roles (Haraway 1997). The scientific practice made possible by modern ways of thinking featured the redefinition of elite English masculinity as "objectivity," while women, working classes, and culturally distanced others were considered hysterics and embellishers incapable of objectively engaging and representing the world at large (Bauman and Briggs 2003; Briggs 2004; Haraway 1997). It is also noteworthy that as modern discourses about gender have traveled to localities formerly colonized by European imperial powers, they have inspired people to fashion novel ways of imagining the gendered self just as much as they have incited local populations (who see this perspective as a hegemonic cultural imposition) to resist (Mohanty, Russo, and Torres 1991). As an alternative, Chandra Mohanty, Ann Russo, and Lourdes Torres (1991) suggest that intellectual and social movements such as feminism must be pluralized and that "third world feminisms" must be articulated from women in specific postcolonial localities in ways that are sensitive to the meaning-laden ways they make and experience gender difference.

In the post–World War II era, Harry Truman and other political leaders further consolidated the connection between development and modernization and gave rise to the idea that many postcolonial nation states in Africa, Asia, and Latin America that were not aligned with the Warsaw Pact countries (the so-called Third World) could be transformed into the sociopolitical equivalents of the United States and Western Europe through the simple transfer of technology, political organization, and integration into global capitalist markets (Escobar 1995, 1997; Ferguson 1999; Isbister 1991). In practice, during the three decades that followed, ideas about development and modernization entertained by national policymakers often carried moralistic assumptions about social normativity that were expected of the "modern" households (and gender relations) of developed nations (Ferguson 1999; Holston 1989; Rabinow 1995).

During the past thirty years, a number of ethnographic studies have demonstrated that envisioning development as the means to obtain a predetermined social, technological, and political order is more a utopian fantasy than an accomplishable objective (Arce and Long 2000; Ferguson 1999; Holston 1989; Rabinow 1995). Counter to the conceptualization of development as a linear process with a knowable endpoint, many contemporary social scientists advocate a view of development as a multi-outcome process of exchange (and sometimes imposition) in which "beneficiaries" actively interpret and refashion technologies, social mores, techniques of governance, and gender relations on the basis of locality-specific meanings, political practices, socialities, and human ecological relationships (Adams 1998; Arce and Long 2000; Gaonkar 2001; Ferguson 1999; Ong and Collier 2005). In naming the outcomes of these processes of exchange, interpretation, and reconfiguration "alternative modernities," Dilip Gaonkar (2001) conveys the idea that modernity, like development, does not have a singular manifestation but is a condition that materializes in a heterogeneous fashion in various parts of the world.

While I find it tempting to use Gaonkar's concept of alternative modernities to name the conditions that manifested in Limón de la Cerca over the course of reconstruction projects, I find the term *postcolonial modernities* to be more apt. *Postcolonial* is a term that various scholars use to convey the idea that although colonialism is considered to have officially ended in Honduras in 1821 with the declaration of New Spain's independence, the implications of colonial social orders and governance practices continue to have reverberations in the present (Chakrabarty 2000; Mbembe 2001; Thurner 2003). Postcolonial, rather than indicating a period after colonialism, is used to denote colonialism's process of evolving ever after (Thurner 2003). As demonstrated in chapter 2,

NGO and local government practices that indexed modernity were deployed in the midst of sociopolitical relationships between disaster survivors, experts, and public officials. The history and logic of these sociopolitical relationships were deeply rooted in Honduras's colonial past and, following independence, in the sequence of imperial relationships that unfolded between this Central American nation and world powers such as England and the United States. In this chapter, I look at the effects on gender of the spatial, material, and social conditions that took shape through these reconstruction practices in Limón de la Cerca.

Even though development is sometimes invoked as a panacea for socially created vulnerability and gender inequity, the critical social science literature on development would have us recognize that development is a concept that takes on varying meanings depending on the institutional contexts in which it is used. Furthermore, the word *development* can stand for practices that mitigate disasters and gender inequity as much as it can stand for practices that enhance them. The case of post-Mitch housing reconstruction programs in southern Honduras provides a window into the messy entanglements of technologies, discourses, governance practices, politics, and subjectivities that go into disaster reconstruction when conducted as accelerated development. In this chapter, I show how gender relations factor into and are transformed over the course of disaster reconstruction programs and call attention to the role of NGOs and local government in this process. Most important, the case of post-Mitch reconstruction in southern Honduras shows how phenomena that are sometimes branded as "global" (i.e., modernization and development) take on locality-specific forms as the myriad social actors—government officials, NGO project managers and workers, architects, engineers, disaster survivors of various socioeconomic backgrounds—who interact in catastrophe-affected communities interpret and reconfigure the practices and techniques of disaster governmentality.

The case of Limón de la Cerca also highlights an important quality of development practice in disaster recovery. As chapter 2 demonstrates, the "state of emergency" temporality that characterizes the way aid agencies approach reconstruction programs results in a prioritization of budgets and fiscal measures over a focus on "cultural" or gender concerns as a means of evaluating projects. While the agencies working in southern Honduras were commonly engaged in development endeavors, their approach to post-disaster reconstruction represents a type of development practice that differs from non-disaster contexts in their acceleration, perceived urgency, and suspension of customary oversight mechanisms. While the vast literature on gender and development has had a significant impact on development practice, post-disaster recovery, I argue, presents unique challenges that remain to be addressed.

Approaching post-disaster reconstruction in a way that prioritizes gender also requires an awareness of the complexities of gendered relations across human communities. First among these complexities is the recognition that gendered ways of being and gender roles vary across time and space, cannot be neatly divided into dichotomies, do not guarantee "in-group" solidarities or homogeneities among women or men, and are not inherent essentialisms. Instead, gendered ways of being are emergent and embodied qualities that are shaped in the midst of ever-fluctuating, meaning-laden relationships among people and between people and things—for example, technologies, environments, spaces (Cupples 2007; Haraway 1997; Latour 1999; Low 2011).

In the case of disasters, a number of studies have shown that there can be significant differences in how women and men in a given community experience catastrophes and in how women and men negotiate and come to embody new subjectivities after a traumatic event (Cupples 2007; Stehlik, Lawrence, and Gray 2000). I find Doña Concepción's reflections on her life in Limón de la Cerca interesting because she chose to speak in affective terms about the shifts in her relationship with her partner (precipitated by changes in her relationships to other women, men, and the built environment of Limón de la Cerca). Things did not *feel* the same. If gender is, first and foremost, a collection of semiotic-material relationships, what experiences of gender change did other residents of Limón undergo, and how did they affectively experience them?

Gender, Space, and Violence

On May 3, 2001, Doña Maria Elvira Maradiaga and her father, Don Carlos, tried to give me some insight on their life in Limón de la Cerca. The conversation took place in their house, which resembled many others in the reconstruction site. The single-room structure was sparsely adorned. Two wooden beds were arranged against the walls to make the most of the available space. One wooden table, which was covered with a plastic tablecloth, provided the center for housekeeping activities. An assortment of plastic dishes, frying pans, and bowls covered its top. In an adjacent small wooden cabinet with glass doors, they kept important documents, medicines, and valued belongings.

Outside, in a temporary structure made of wooden beams and a plastic tarp, two young women helped Doña Maria Elvira make tortillas for sale in the town market. As he spoke, Don Carlos Maradiaga, a man in his sixties, sat on the bare concrete floor of the house. He seemed physically and emotionally exhausted. In response to a question about their appraisal of their lives in the reconstruction site, Doña Maria Elvira said their situation was critical. When asked why she thought this was the case, she replied:

MARIA ELVIRA MARADIAGA (MEM): Because here, everything . . . you have to see how you're going to eat, and it is so much work to live day to day, and sometimes work doesn't work out! Because I sell tortillas, and there are days when I come back with [makes gesture of large pile] of tortillas. I work hard here, not like when we were in Choluteca. Even for firewood . . .

CARLOS MARADIAGA (CM): And it is a long way.

MEM: How long it is, here we have to pay bus fare, and at least over there in Choluteca, we were close to the market. We went on foot, and not here, and here you can't even wake up before dawn. We are afraid.

ROBERTO BARRIOS (RB): Why are you afraid?

MEM: So much *mara* [gangs] that there is here.

RB: Then when it's dark you don't . . .

MEM: Well, look, there's no light! [Electricity was not yet installed in the resettlement site, a punishment from the mayor for the disaster survivors' political dissent.]

RB: How is the power situation? Have they told you when they'll put it in?

MEM: Nah.

RB: They said it would be done by November . . .

MEM: No, look at how the sewage project is going [another stalled infrastructure project]. How must the electricity project be?

CM: Look, because here, even this daughter of mine, she gets up, she wants to work, she wants to work, but alone she can't, she can't. She has to look for someone here in the house so they can go in the night, in the darkness to the mill. Alone, she can't go, because they could come out, because here the *maras* hang out. You should see how they go—yes, there. Then they can take the maize bucket, they can dump it, and, well, they can hurt her, and she, by herself, she can't . . . Then those things one sees.

(structured interview transcription 2001)

Like Doña Concepción Rodriguez, Doña Maria Elvira and Carlos Maradiaga noted that the resettlement site was a place where novel social and spatial arrangements among residents and between residents and things were taking shape. The random distribution of land parcels created conditions of social fragmentation in which adolescent gangs could thrive and act with impunity, and the Maradiagas experienced these conditions affectively: "We are afraid."

Unlike Doña Concepción, however, Doña Maria Elvira continued to work outside the home, selling tortillas in the town market. In the case of the Maradiaga household, Doña Maria Elvira—who instead of living with a male partner resided in an extended household with her father, teenage brothers, and her young children—was one of the primary sources of income. Although she continued to be involved in her pre-Mitch occupation, Doña Maria Elvira noted that spatial and social conditions were such that she had to work harder and face the threat of physical violence to make a living. As our conversation continued, Don Carlos Maradiaga told me just how real that threat was:

Because look, one time, one of them, right here, they gave me this big story that she had gone to report them to the police, those who were in the *mara*. And she wouldn't even go out because she was scared. Well, they came here. Look, some people, and such men, to question her. Then she said, "And what am I going to do, getting involved, going to report you, when I don't even know you?" They told her, "But you are the one who went, since there is proof, look!" That is what happens. Then they came and whipped her. Then that is how that is. One in his home, and you think you are safe, and they said, "If we knew for certain who that person was, we would burn them alive." Imagine, like that, all those things give you fear. (structured interview transcription 2001)

Don Maradiaga's story shows that local government practices transformed social and spatial relations in the resettlement site, but instead of engendering a modernity that brought about gender equity, they created conditions that exacerbated the disaster's gendered effects. Moreover, if we consider that gender involves not only women but also relationships among men and between women and men, then Don Maradiaga's words suggest the reconstruction process had transformative effects on the masculinities of displaced Cholutecans as well.

Over the course of my fieldwork in southern Honduras, I resided in Cholutecan *barrios de clase obrera* that were comparable to those of Limón's residents before the storm. In these neighborhoods, male residents often collaborated with each other to protect their households and relatives from gang activity. On one occasion, for example, while I was staying the night at a house in Barrio La Libertad, a car alarm went off in the middle of the night, and the adult men of three households stepped outside with machetes to make sure a robbery was not taking place. These men were not strangers to one another. They had been neighbors for several years and had developed *confianza* through daily socializing. In Limón de la Cerca, in contrast, residents reported not having similar trust with their neighbors to respond collectively to perceived threats.

Just as social relationships among women were disrupted by Mitch and resettlement, with the effect of limiting their access to child care, relationships among men were significantly affected as well. Although Don Carlos Maradiaga was not as explicit in articulating his experience of gender change as Doña Concepción Rodriguez was, the tone of his story certainly conveyed a similar (although differently gendered) sentiment. His gendered role as his daughter's protector, which was a common narrative about masculinity in Choluteca, was undermined by the impunity of *mara* members, and he experienced the reconstruction site as a space of terror.

Gender, Space, and Kinship

Before the hurricane, Doña Dania Lisette Giacoletti and her husband, Don Reynaldo Alberto, lived in Colonia Victor Manuel Argeñal of Choluteca. At the time of our interview, they had resettled with three of their four young children to Colonia Nueva Jerusalén, one of the nine subdivisions of Limón de la Cerca. Unfortunately, the Giacoletti family had not benefited from any of the housing construction programs, and despite the presence of three hundred vacant houses in the resettlement site, they continued to live in a temporary shelter donated by the International Organization for Migration.

Even though they lived in a precarious housing circumstance, Doña Dania Lisette and Don Reynaldo Alberto seemed to have made the most of their situation. They meticulously maintained the land parcel where their *micro* was located, erecting a barbwire fence to mark the parcel's perimeter and attaching a makeshift kitchen to one of the shelter's sides. Inside, the temporary shelter was neatly organized. A full-size bed with an iron frame occupied most of the space in the single-room home. The remaining space was filled by a wooden cabinet, two small chairs for the children to sit on, two small tables for cooking and keeping dishes, and a few posters and calendars that hung on the walls.

On April 18, 2001, I sat down with Doña Dania Lissette (DL in transcription) to speak about her life since the hurricane.

> DL: Well, look, my life, I feel that . . . that it changes a little for the better, right? Compared to before. Now, well, since I only had my three eldest children, then came my son, the little one, the one that's sleeping . . . But it has been a suffering here, being here in these strong temperatures, because we don't even have trees. But we're getting better because, with time . . . We have already planted a few trees over there, that we are already feeling a little bit of breeze because of that, you see.
>
> But in the economic situation also, because you find yourself

uncomfortable, in going to look for work so far away. Because my husband had to go to work for more than half a day, just for a while. Because he couldn't leave me here alone either, with the children, but now I am getting used to it, and, well, we don't live good nor bad, but regular, a tranquil life, yes. Thanks for the food of our children, of the family, right? But there we are, passing by, thanks to God . . .

RB: You told me that your husband didn't like being gone for more than half a day. Why didn't he like to?

DL: It's because, look, his work. It's that he was going around in a taxi . . . The owner of the taxi, to help him so that he wouldn't lose work, he let him borrow the taxi for the half day. So he gave it to him from noon until ten at night. Sometimes he missed the last bus, and sometimes he did catch it. Sometimes I had to sleep alone with fear, and everything, but I was alone with my kids.

Back then, oh yes, but then, later, he got another taxi, that the lady did let him have it since the morning, since breakfast, until six in the afternoon. Then, like that, we didn't have problems. We started getting better because he worked all day with no problem, without thinking about it, and he came to sleep here with us, all night.

That's how he's in the same work. Even if we live day to day, we are making it, with tranquility. (structured interview transcription 2001)

Although Doña Dania Lisette began our conversation in an optimistic tone, she was quick to point out the difficulties the resettlement site posed for her family. As did other residents, she noted the distance to Choluteca as a major impediment in her husband's work habits. In addition, Doña Dania Lisette mentioned the insecurity she and her children felt on those nights Don Reynaldo Alberto was not able to return home in a timely manner. She also noted that the distance between the resettlement site and the city also had an impact on her kin relations, as it separated her from her mother and brothers who had provided her with significant emotional and economic support before the storm:

Yes, it's complicated for me because of how far away I am from my family—from my mother, from my brothers, from everyone. Because it's only me and a cousin who live here, who are related. (structured interview transcription 2001)

Doña Dania Lisette's words resonate with theorizations of gender as a product of relationships among women, among men, and between women and men, but her comments also lead us to consider the role of space in these relationships. She tells us that her proximity to her mother and brothers was critical to her quality of life before the storm, and she proceeds to tell us how her kin relations played a role in shaping the way she thought about herself before and after the storm:

Yes, we're moving forward, asking God to help us and to guide us to raise our children. I'm only twenty-five, and I had my children one following the other. Look at my oldest child, he's only six years old; look, and she's four; and the other one, three. Oh well, look, I was having my children one after another, because the other child is also one year old now, and well, my mother and my brothers saw our situation, of myself with the father of my children. We lived with problems, that we didn't have money. And, well, if it wasn't one thing, it was another.

Then, they gave me some advice, that I get operated [tubal ligation], or, if not, I was having more and more children. So then, the situation, it wasn't a stable life. It wasn't a nice life for my children who were going to be born, then . . . No, I had the operation. I made the decision also because he, he was very, it's that he was, in those taxis, at work, at work he was a womanizer. I was pregnant, really pregnant, and then he, he would distance himself. Then we would leave each other. (structured interview transcription 2001)

Doña Dania Lisette shows how the spatial shifts created by reconstruction practices on the part of grassroots organizers, local government, and NGO housing programs altered her relationships with important women and men in her life. What is more, these changes transformed the way she thinks about herself. Before the storm, her brothers and family were spatially proximate and provided her with a safety net, but life in the reconstruction site weakened these ties. Back in Choluteca, Doña Dania Lisette's extended kin group mitigated Don Reynaldo Alberto's infidelity during her pregnancy. In the resettlement site, in contrast, she became increasingly dependent on her husband. As she explained, the resettlement site created a context of social vulnerability for her, reducing the number of people she could count on for emotional and financial support to one, Don Reynaldo Alberto. This situation was the final factor in her decision to undergo a tubal ligation, an action that shows how the spatial and social conditions of the resettlement site transformed the way she thought about her body and her relationship toward her marital partner.

What is most interesting about Doña Dania Lisette's story is that Limón de la Cerca (with its modernist spatial design) stood as an icon of development in local government brochures (fig. 9). Nevertheless, this development project did not necessarily enhance women's agency and gender equity; rather, it created conditions of alienation from extended kin networks and of dependency for many women. Women like Doña Dania Lisette were not completely deprived of agency in this situation. Her decision to undergo a tubal ligation demonstrates that she was an active agent in her world, taking steps to improve her life and that of her family. However, this did not mean that the development projects of Limón automatically created progressive improvements in her household's gender relations. Instead, these projects presented her with a collection of social and spatial complications that she resolved with available resources and within the discursive possibilities laid out in her kin relations (i.e., the advice of her brothers and mother to have a tubal ligation).

Transgender, Space, and Isolation

Discussions of gender in post-disaster recovery often focus on people who neatly fit what are widely accepted gender categories, and up to this point, this chapter has continued this tendency. My discussion of gender in post-Mitch Choluteca and Honduras, however, would not be complete without mentioning transgender individuals who had a visible presence not only in Limón de la Cerca but in Honduras's major cities as well. Indeed, well-known ethnographies of gender in Latin America often focus on the meanings of masculinity (Gutmann 1996) or the experiences of women (McClaurin 1996), but transgendered Central Americans are easily overlooked in these analyses. Meanwhile, as someone whose field-work focuses on Central America, I was actually impressed with the public presence of transgendered Hondurans even in Choluteca, which many Hondurans considered provincial and culturally conservative.

The most visible instances of transgendering involved people whose physiology is described as male in the categories of biological science and who wore attire and practiced bodily gestures meant to index femininity in Honduras. Despite their notable presence in the city's urban landscape, transgendered individuals faced socially imposed limitations in the kinds of physical and social spaces they could inhabit both before and after the storm. Most interesting, the spaces where they were visibly present were also those most closely associated with normative masculinity in Choluteca, such as the movie theater and the bars and pool halls surrounding the central market and bus terminal.

The movie theater, for example, featured a spatial code—that is, in a given cultural context, practices and behaviors that are considered appropriate in a particular space—that was quite different from what I had become accustomed to during my adolescence in suburban New Orleans. In Choluteca, the movie theater was where adult men socialized, smoked cigarettes, and engaged in romantic adventures. Cholutecans I spoke with, for example, considered it inappropriate for single women to visit the theater by themselves, and teenagers with disposable income, who were rare in the city to begin with, were generally absent. On one occasion, when I visited the theater with Rosa Palencia, a transgendered Cholutecan walked by us on repeated occasions and politely asked me to move my legs so she could walk down the aisle. After the second of these instances, Rosa jokingly teased me, "I think she likes you—that's why she keeps asking to pass in front of us" (unstructured interview 2000).

Rosa's teasing opened up a conversation about transgendering, gender relations, and that word used to represent and homogenize the masculinities of Latin American men, *machismo*. The next day, while riding in a taxi from Choluteca to Limón de la Cerca, I confided in Rosa that I was impressed by the high visibility of transgendered people in what I would have considered a culturally conservative region, especially with regard to gender norms. With a mischievous smile, Rosa responded:

They like it. Women are expected to be submissive, but the transvestite can be both feminine and sexually aggressive, and I think men like that. (unstructured interview 2000)

Then Rosa quietly gestured we should postpone our conversation, as our eavesdropping taxi driver was becoming increasingly irate as our conversation developed. He was angry with me for asking such questions and with Rosa for speaking about a topic that was off-limits.

In Limón de la Cerca, only two of the roughly nine hundred inhabited households were home to openly transgendered individuals. One belonged to Don Julio Salazar, a resident who engaged in transgendering activities outside the community but who assumed a male gender role while in Limón. I did not intend to survey Don Julio's house as part of my research because its number had not come up in my initial randomized selection of households. Nevertheless, Don Julio was determined that I would not pass him by; he wanted to become part of my ethnographic project. On an overcast but hot day in May 2001, Don Julio called to me from his front door as I made my way to interview the residents of another house in Limón. When I answered his call, Julio invited me into his house, which was sparsely furnished, even when compared to the austere interior of other disaster survivors' homes. Don Julio offered me a seat on a plastic lawn chair and sat on his bed, which occupied most of the small, single-room structure.

One of Don Julio's motivations for inviting me in was to find out what the "gringo in the cowboy hat" (I had taken to wearing a Stetson to protect my face from the harsh Choluteca sun) was doing walking around the reconstruction site. Although I had originally explained my research to local authorities, public health workers, and community organizers, informing the entire population of Limón de la Cerca had proven a daunting task. Consequently, I became accustomed to taking the time to speak with people and families who, although not randomly selected for the survey, were curious about my work and its implications for the site's residents. After I explained my project, Julio and I talked about his life in Limón, and he was quick to tell me about the recent demise of his relationship with his live-in partner. Don Julio seemed lonely and isolated. Unlike other housing structures that were inhabited by multiple generations and extended family members, the sparsely adorned interior of the house demonstrated that he was the only person living in it. After some pleasantries we said good-bye, and I went on my way to speak with his neighbors.

Two days later, while I checked in at the Limón de la Cerca health center, one of the nurses I had befriended, Luisa Martinez, called me aside and said she had heard of my visit to Don Julio's house. When I confirmed we had spoken, she commented: "Well, be careful, because he has AIDS" (fieldnotes 2001). This brief conversation served as a reminder that despite the conditions of relative anonymity in Limón, my behaviors and those of other residents were closely monitored. Furthermore, I found Luisa's comment intriguing. As a member of the clinic's staff, she had access to Julio's health history; I had seen him visit the clinic on multiple occasions. As a state health worker, Luisa was officially required to safeguard the privacy of her patients, but she had chosen to disregard this legal requirement. Following our brief exchange, I was left to wonder whether Luisa had shared this information with me because she had a genuine—although poorly informed—concern for my well-being (HIV cannot be contracted through casual contact) or whether she was collaborating with Don Julio's neighbors to police his and my gendered identity and sexuality. Perhaps Don Julio was not even HIV positive at all. Was the comment meant to keep me from socializing with him? I would never know the answer.

The point of this ethnographic vignette is that transgendered people are part of disaster-affected communities—even if they are a small minority, as they were in Limón de la Cerca. Furthermore, their ambiguous and arbitrarily imposed status as "fetish" (based on Rosa's claim that their simultaneous play on masculinity and femininity made them sexually desirable to many men) and "deviant" (as implied by Luisa's implicit suggestion that I not associate with Julio) leads to unique complications for transgendered people during disaster recovery that must be accounted for in gender analyses.

Tom Boellstorf (2005, 2015) has made the case that gay, lesbian, and transgendered people (LGBT) face unique spatial complications. The making of modernist national spaces, he argues, often involves the simultaneous production of hetero-normative identities and affects. As an effect of these nation-building practices, LGBT people must make social and material spaces for themselves, a process that is often contested and resisted by those not sympathetic to their cause. This observation must not go overlooked in community relocation programs. While heterosexual and non-transgendered women and men experienced difficult shifts in gender relations over the course of recovery in Limón, they still enjoyed the ability to make claims to social normativity, whereas transgendered individuals faced the compounded challenge of wrestling with both their status of social liminality (which already denies them a social space) and the dramatic transformation to material spaces and social relations caused by disaster recovery.

Conclusion

The ethnographic interview excerpts discussed in this chapter show how the reconstruction practices of grassroots organizers (site selection), local government officials and engineers (random land distribution, modernist regimentation of space in the site's master plan), and NGO housing program managers and architects (minimal and haphazardly constructed housing units) created spatial and social conditions that presented unnecessary hardships for disaster survivors: increased violence, anonymity, alienation, lengthened distances to work sites, and difficulties with accessing child care. These hardships had transformative effects on gender relations and enhanced gender inequities in the reconstruction site. What I find interesting is that the practices of local government officials and NGO program managers routinely articulated inherent modernist and developmentalist assumptions about the natures of people, communities, and social well-being. Furthermore, this case study demonstrates that techniques of "modern governance" on the part of local government and NGOs were often intimately entangled with practices of postcolonial political culture—that is, clientelism and the use of reconstruction resources to reward or punish docile and assertive grassroots organizers.

The unique kind of modernity fashioned in Limón de la Cerca did not produce the gender equity often envisioned as the objective of development programs. While the conditions and experiences narrated by men and women in Limón could be described as a reinstatement of "traditional" gender roles (e.g., relegating women to the domestic sphere), I emphasize that there was nothing traditional in the context of southern Honduras about the enhanced inequities that developed over the course of Choluteca's reconstruction. Rather, the conditions documented in this chapter are very much an effect of the particular kind of postcolonial modernity engendered through local government and NGO reconstruction practices in Limón.

The case of Limón de la Cerca highlights the importance of approaching disaster survivors as members of broader, gendered social and spatial relations. Those relationships involve people as much as they involve things like space and houses. Decisions on the part of local government and NGOs in Limón routinely treated displaced Cholutecans as alienated beings, or individuals who could be easily spatially rearranged through the random distribution of minimal land. Disaster survivors, in turn, proved to experience their gendered sense of self in the midst of long-term relationships with their neighbors and the members of their extended kinship groups. These relationships featured an important spatiotemporal component, as proximity to longtime neighbors and to family members played a key role in building sentiments of trust and familiarity. Once established, these sentiments provided a social ground for facilitating child care, home security, and a sense of comfort.

Disaster reconstruction policymakers must recognize the importance of gendered social and spatial relations and approach reconstruction practice not simply as building infrastructure but also as reestablishing affectively experienced social lives. Policymakers should also be leery of reconstruction designs that seek to simplistically modernize disaster-affected populations. Instead, they should strive to identify the ways disaster survivors speak about the relationships among people and between people and things that make life meaningful and intelligible in their communities. Disaster survivors seldom speak with one voice, of course, and gendered differences in the ways people make social relations and space are certain to emerge through their narratives. Nevertheless, if we aspire to devise recovery programs that address gender inequities in locally relevant terms, then it is essential that we ethnographically apprehend, appreciate, and wrestle with the multi-vocality of disaster survivors.

4. The *Marero*

TERROR AND DISGUST IN THE AFTERMATH OF MITCH

My earliest recollection of hearing the term *marero* dates to 1986, when I was twelve years old and living in the southern suburbs of Guatemala City. Classmates and older teenagers used the term to refer to participants in what they considered an emerging social phenomenon—adolescent street gangs involved in armed robberies and territorial vendettas in the city's central areas. My acquaintances and other Central Americans derived the word *marero* from the colloquialism *la mara*, a term used to speak of one's close group of friends that is rooted in the Spanish word *marabunta* (a swarm of insects).

As a relatively sheltered child of educators, I never saw or met a *marero* before my family migrated to the United States in 1987. *Mareros* were more a figure of urban lore than an actual group of people with names, faces, and histories. Middle-class Guatemalans used to playfully terrify each other over casual conversations, telling stories about the gang members' gruesome acts of violence. I did not return to Central America until 1999, when I traveled to Honduras to conduct preliminary research for my dissertation fieldwork. As I visited the temporary shelters housing hurricane-displaced families, I noticed the ubiquitous presence of a peculiar form of graffiti—the spray-painted image of a hand with long, claw-like fingernails and its index and little fingers pointing upward. I was accustomed to associating this gesture with innocuous North American heavy metal music culture. Here, however, the devilish hand was often accompanied by the number 13 and the acronym LMLS, which stands for La Mara Loca Salvatrucha (the Crazy Salvatrucha Gang). The graffiti appeared on newly built houses constructed by international relief agencies and on NGO billboards that greeted visitors to resettlement communities. As it turned out, the *mara* phenomenon had grown significantly in twelve years. The Salvatruchas were one of two major, loosely affiliated transnational gangs operating in Guatemala, El Salvador, Honduras, Nicaragua, and the United States.

These gangs developed during the Cold War, when transnational migration to North American cities such as Los Angeles exposed displaced Central Americans to U.S. gang culture. Socially marginalized Latin American youths, following their deportations from the United States, then introduced their newly acquired way of imagining identity to Honduras, El Salvador, and Nicaragua. The successful U.S. Cold War intervention and state government policies in Guatemala, El Salvador, and Honduras designed to eradicate labor and leftist movements (which called for addressing social inequity) during the 1970s to the 1990s left few venues for disenfranchised youth to express dissent and develop empowering identities (Way 2012). In this vacuum of possibilities for counter-identifications, *maras* found a fertile ground.

Anthropologists of disaster are often interested in social organizations that emerge in the aftermath of catastrophic events (Fortun 2001; Oliver-Smith 1999). These emergent organizations usually form when members of a community (imagined, multi-local, virtual, or geographically circumscribed) come together in response to pressing circumstances that existing governmental institutions or NGOs are not addressing. In the case of Limón de la Cerca, the local chapter of the Mara Salvatrucha was the most prominent emergent organization after Hurricane Mitch. As discussed in chapters 2 and 3, aid distribution practices and the local government's suppression of the displaced Cholutecans' grassroots leadership effectually fragmented social networks and denied the nascent community its most proactive members. The *mara* quickly filled the resultant vacuum in community organizational structures.

In Choluteca the local police department claimed a tenfold increase in gang membership following the catastrophe, roughly estimating five hundred youths and young adults were involved in such organizations. At a national level, estimates of *mara* activity claimed up to three hundred gangs, with roughly thirty-five thousand members, existed throughout the country's urban areas (Mencias 2002). Some *maras* were small, with only a dozen participants, whose ages ranged from childhood to adolescence. The members of these smaller groups usually engaged in panhandling, scavenging, petty theft, and consuming narcotic inhalants.

Larger *maras* were credited with higher membership numbers, with people ranging in age from adolescence to adulthood, and were rumored to include former military servicemen in their ranks. The activities of these larger groups included extorting "protection" money from public transportation drivers, conducting armed robberies, kidnapping, and murdering rival gang members and police informants (Mencias 2002; Thompson 2004).

By 2000 when I returned to conduct my long-term fieldwork in Honduras, *maras* were everywhere and not just in the form of young men and women with their shaved heads, baggy clothes, tattooed bodies, and graffiti tags but also in the form of news stories, urban myths, paramilitary counter-gang campaigns, and Hondurans' everyday conversations. In the city's coffee shops, people discussed what they considered available options for dealing with the *mara* "problem": state-sponsored violence, domestic disciplining, or vigilante "justice." Honduran social life had become *mara* saturated. It was in the midst of this wave of sensationalism that I sat at an Internet café in Tegucigalpa and, while waiting for a computer terminal to become available, read a news story about *mareros*.

The story went something like this: In San Pedro Sula (Honduras's industrial urban center), a group of Salvatruchas and some members of their rival gang, the Mara Dieciocho, were forced to share a warehouse as their sleeping quarters in prison. A fight broke out between the two gangs, and eleven or so Salvatruchas were killed during the brawl. A few weeks later, prison guards found chopped fragments of human bone in the sewage (*Diario el Heraldo* 2000a; *Diario la Tribuna* 2000). After conducting a head count, the guards determined that a high-ranking Salvatrucha member was missing and speculated he was killed, cut up, and eaten by his fellow gang members for his poor performance during the fight. The printed story, spread over a two-page centerfold, featured a large photograph of a *marero*'s back adorned with a tattoo of the gang's logo. As I read the text around the photo, I had difficulty discerning what had actually transpired in that communal jail, and the story seemed to flirt with the lines separating journalism and fantasy. The only thing I could be certain of was that my body did not remain passive as I imagined the described events. A chill went down my spine, an ambiguous sensation that was as terrifying as it was pleasurable, or what Brian Massumi (2002) would call the indeterminacy of affect.

In the years that followed, I grew accustomed to seeing newspaper headlines that detailed violent acts attributed to *mareros* as well as the extrajudicial execution of suspected gang members by paramilitary groups and other unnamed assailants. Following September 11, 2001, *mareros* even came to be seen as a potential terrorism threat in Central America and the United States. The January 2005 issue of the Jamestown Foundation's *Terrorism Monitor*, for example, included an article titled "Al-Qaeda's Unlikely Allies in Central America," which read: "Recently there have been reported sightings of al-Qaeda operatives in Honduras. According to some observers, their alleged presence in that country conforms to their desire to secure land routes to the United States, through collaboration with Central American gangs" (Pineda Cruz 2005, 3).

Reports from the office of Honduran president Ricardo Maduro also painted a macabre and terrifying image of the street gangs. Detailing the events of an anti-*mara* operation, an online document read: "In the operations, which began sharply at five in the morning, the police confiscated a coffin as evidence that the gang members practice satanic rituals, an AK-47 rifle, a rocket launcher, a cellular phone, one beeper, two computer diskettes, and one photographic roll among other objects" (Casa Presidencial de Honduras, electronic document, 2003).

Beyond digital media, as noted, *maras* also became the subject of urban lore. One particular story that circulated among Hondurans I spoke with concerned the gangs' initiation rituals, which allegedly required prospective male recruits to kill a member of a rival group and prospective female recruits to have sex with all gang members. The depictions of Honduran gangs in these news stories and urban legends had the potential to create what Jean Baudrillard (1995) has termed the *hyperreal* or what Michael Taussig (1993) has described as the mimetic "power of the copy" over what it is supposed to represent. What Baudrillard and Taussig mean is that a representation is never an objective rendering of that which it purports to describe. Instead, it is a culturally situated interpretation of its subject. Moreover, the representation can become "more real" for its readers, listeners, and observers than the people it claims to document, making the latter subject to judgments based on the secondary reality created by the image. In other words, the represented is held up to a standard extant only in the representation.

As a consumer of news media and urban lore, I had no means of verifying whether incarcerated Hondurans actually ate their fellow *mareros*, although I did consult a *Time* magazine investigator who verified the prison brawl deaths but doubted the veracity of the cannibalism story. Nor could I confirm whether al-Qaeda operatives were indeed present in Central America and were actually establishing connections with youth gangs. But perhaps verifying the truth of these stories was not as anthropologically interesting as trying to ethnographically trace how these stories circulated through Honduran society and how their readers and listeners experienced them at an affective level—that is, what their bodies did as they imagined the described events, what moral positions their bodily reactions to the stories led them to take, and what actions they actually took as an effect of *mara* hyperreality.

Collectively, the stories about *maras* seemed to create an image of *mareros* as an imminent external threat to the Honduran—and perhaps even North American—populace. Representations of gang members figured them as violent satanic cannibals, as subjects outside all civility, and as a fetishized threat to Central American society. One thing that concerned me about these representations was the way they ignored the role of colonial history and twentieth-century U.S.–Central American relations in creating the conditions of possibility for the *mara* phenomenon in the first place (Way 2012).

Sensationalized accounts of gang violence, I felt, urged readers to take positions that engaged the issue symptomatically. The site of the "problem" was the *marero* itself—a subject that had to be interdicted, reformed, or destroyed to protect the common good—and not the state boundary–crossing, sociopolitical relations and the colonial and postcolonial conditions of inequity that engendered it. These perspectives on the *mara* phenomenon manifested in the political campaigns of the first Honduran presidential election following Hurricane Mitch. At the national soccer stadium in Tegucigalpa, massive loudspeakers blared campaign advertisements endorsing Ricardo Maduro's bid for the presidency: "Con Ricardo Maduro, su futuro está seguro!" (With Ricardo Maduro, your future is secure!) The slogan was preceded by a reminder to listeners of Maduro's extensive experience as a businessman in the private sector and as president of the Central Bank of Honduras. In Maduro's campaign, the idea of security intertwined promises of economic prosperity with the violent eradication of the *maras*. Many of the Hondurans I spoke with about the ads interpreted them as a pledge on Maduro's part to institute *mano dura* (hard hand) crackdowns and exterminations of *mareros* by police and paramilitary groups.

While in previous chapters I focus on how the affective experience of recovery on the part of disaster survivors is often missed in the policies and practices of NGOs and local governments, this chapter takes a different approach by examining the role of emotions in the operation of state power in postcolonial societies. This line of inquiry builds on Brian Massumi's (2010) explorations of the ways in which the sensation of being threatened is used in contemporary security states to discipline populations and to create subjects who behave in predictable and collective ways. This analytical thread also concurs with the observation that power is not possessed or hoarded by a single social actor who then imposes his or her will on a population; instead, power works through culturally cultivated bodies who desire and fear in particular ways and consequently lead people to act collectively toward a particular end (Deleuze and Guattari 2009; Foucault 1978; Stoler 1995). At the same time, the ethnographic work upon which this chapter is based requires us to take such analyses one step further and to consider how the affect-laden social and kin relations among subaltern Hondurans curtail the power of hegemonic, terror-inciting narratives about *mareros*.

The case of Honduran *maras* provides a unique setting for exploring the ways power operates in the postcolonial context of Central American nation-states. If media stories about *mareros* had the effect of producing both pleasure and terror in my body as I read them, what were other Honduran readers experiencing? What kinds of moral positions were they taking? Furthermore, if bodies and affect are shaped through experience and if, as Judith Farquhar (2002) has argued, reading a culturally structured narrative such as a novel or news story is an experience that can shape the way we experience future incidents, then what were these stories about *mareros* producing? What kinds of emoting subjects did these stories bring to life? Furthermore, what were the limits of the kind of power articulated by *mara* stories? In what ways was this power contested, and what other factors were involved in shaping emoting bodies in Honduras?

The Ethnography of Maras

Taking an ethnographic interest in *maras* required me to address a number of methodological complications. Obtaining access to *mareros* and conducting participant observation among them in Limón de la Cerca, although seemingly difficult, were not as problematic as negotiating the relationships I would establish in the process. My Cholutecan research assistants, who were not in *maras*, warned that *mareros* who agreed to become interlocutors would develop expectations of loyalty on my part to their gang, and that could put me in the middle of potentially dangerous interactions with the police and other gangs. Reconstruction site residents who were not *mareros*, I was also advised, could resent me for associating with people who assaulted or threatened them, and that would interfere with my research on disaster recovery. I decided to handle these methodological complications by relying on a multi-vocal and multisited methodology and by limiting my engagements with *mareros* in the reconstruction site to observing their interactions with other residents in public settings like buses and community festivities. During the course of my ethnographic research, I also conducted observations in more private settings, such as the dining rooms of families I interviewed who had *mara*-involved children or siblings. I also sought contact with *mareros* away from the reconstruction site, where I felt my association with adolescent gangs would not have the consequence of arousing suspicion or resentment from non-gang members. In the city of Tegucigalpa, I relied on the assistance of Leah Getchell, a Covenant House volunteer, to visit groups of homeless adolescents who identified as *maras*.

Because my research questions were concerned with the hyperreality of the *marero*—that is, the *marero* as he or she exists in newspapers, websites, and stories told by people of various social positions in Honduras—my work also included the collection of gang representations in mainstream Honduran media such as newspapers and state websites. I complemented my analysis of these texts with observations of people reacting to them. In the course of my ethnographic research, Internet cafés in Tegucigalpa became useful sites where I could observe people consuming *mara*-related media and discussing these organizations. In early twenty-first-century Central America, where the lower-middle and working classes do not have the discretionary income of their North American counterparts, Internet cafés have appeared as both viable businesses and new localities of socialization where one can observe and participate in conversations about current topics.

The Historicity of Terror and Disgust in Honduras

Sitting at an Internet cafe in Tegucigalpa in 2000, nineteen-year-old Cecilia Rodriguez decided to tell me why young Hondurans regarded joining a *mara* as a reasonable thing to do:

> La gente siempre los ha visto como mierda, Pero para ellos vos sos la mierda (People have always seen them as if they were shit, but for them *you* are the piece of shit). (unstructured interview 2000)

Cecilia resided in a working-class neighborhood in the Honduran capital city, and she knew several members of her generation who had joined the ranks of La Mara Dieciocho. Her father was an absent transnational migrant worker who had spent the last six years in various cities in the United States, so Cecilia empathized with gang members even though she had never participated in the organizations. As the daughter of a family of very modest means, her employment prospects were limited to working as a sales clerk, a job she had once tried and found unbearably dull, in the nearby business district of Comayagüela. As an alternative, Cecilia had enrolled in a school for artists and earned a minimal amount of money from the dance and theater performances that the school organized for the public.

By speaking about how an indeterminate observer sees the adolescents who joined *maras* as fecal matter, Cecilia summed up her take on the affective dimensions of class difference in Honduras. Class difference, she suggested, is more than just a difference in wealth, labor, cultivation, and behavior; it is experienced through bodily sensations such as disgust or pleasure, sensations that are themselves linked to moral and ontological judgments. Elite and upper-middle-class Hondurans, she claimed, thought and felt the youth of urban poor and working classes *were* disgusting shit.

I found Cecilia's words intriguing because they also succinctly brought together affective dimensions of class difference and gang identity involved in the *mara* phenomenon. Joining a *mara*, her words indicated, allowed a person to reverse a hierarchical social order that placed the members of Honduras's working class in a subordinated position. Cecilia's words resonate with the observations of Ranajit Guha (1983), one of the founding figures of subaltern studies, who wrote about another group of people whose actions, like those of *mareros*, were once represented as illogical antisocial aberrations—Indian "peasants."

In *Elementary Aspects of Peasant Insurgency in Colonial India* (1983), Guha took issue with the colonial bias reflected in how many historians of South Asia represented peasant revolts as apolitical and irrational behavior of mindless mobs. Such histories, Guha argued, held the colonial government as their implicit subject. They assumed this form of governance and its accompanying social arrangements of ethnic, race, and class hierarchies and their techniques of imperial violence were unquestionable norms and necessities, but these histories failed to recognize the government's role in giving form and reason—that is, the feeling of injustice and inequity—to the phenomena in question (i.e., *maras*, "peasant" revolts). For Guha, the actions of Indian insurgents—for example, destroying debt records and specific elite households, reversing roles with elites, subverting linguistic patterns meant to index caste and social standing—indicated that peasant rebellions systematically undermined those practices and signs that structured colonial societal order in India and enabled the subordination and economic exploitation of colonized populations. Tracing the actions of insurgents, then, allowed one to map the technologies, techniques, symbols, and cultural practices that determined who could occupy what social, institutional, and geographic spaces and who could do what labor.

As I have continued to think about Cecilia's words, I have found Guha's work helpful, because it suggests that the actions of Honduran *mareros* reveal a schematic of the way power relations are arranged in Central American nation-states. In such a light, one could interpret the seemingly irrational behaviors of *mareros* as an attempt to reverse the signs and practices through which power is structured in Honduran society and, with such action, end their subordinated status as marginalized members of the urban working poor. But what were these actions, what signs did *mareros* attempt to reverse or obliterate, and how was affect implicated in the process?

After posting their pervasive graffiti, the human body seemed to be the second site upon which *mareros* did their work. They tattooed their hands, faces, necks, and backs. Some of these tattoos were meant to communicate what they had done to the bodies of rival *maras*' members. In Limón de la Cerca, Erica Palacios, a twenty-two-year-old college student and displaced Cholutecan, explained the meaning of three crosses tattooed on her brother's back. "Each one represents a Mara Dieciocho member he killed" (unstructured interview 2001), she told me during a conversation about her brothers. Shaved heads and over-sized baggy shirts and pants also communicated *mara* membership.

For Michel Foucault, this emphasis on the body—and what seemed to many non-*mara* members as its defacement—would not have been insignificant. In *History of Sexuality* (1978), Foucault made the case that nineteenth-century bourgeois class formation centered around the cultivation of specific kinds of bodies and dispositions; disciplined bourgeois bodies were thought to manifest biological optimization and universally desirable ways of being, even though such bodies were the materialization of culturally particular conventions. As Foucault put it, the bourgeoisie had to make a body for itself, and this body was both the site from which this class operated as a hegemonic group as well as its product. Following this line of inquiry, Ann Stoler (1995) has suggested that making the bourgeois body also required cultivating desire, as the desire for this way of being was the driving mechanism of power in class formation. From a foucauldian perspective, power does not oppress all of its participants; rather, it incites and moves people to particular actions and ends. The body of the bourgeoisie, in turn, was tied to the emergence of what Foucault called biopolitical governance—that is, a state whose object of concern was to care for people as living organisms and where fostering biological life came to take precedence over other kinds of social life.

The Honduran state, as with many a postcolonial locality (see Gupta

2012), is far from actually being an effective collection of biopolitical institutions. Instead, as I detail in chapter 2, the state is the product of a colonial history. Its political culture took form as a means of securing wealth and influence among the caudillo and *latifundista* (owners of large landholdings, often accumulated through the dispossession of subsistence farmers) classes, and this political culture received new life under U.S. influence during the twentieth century. Thus, the Honduran state only mobilizes biopolitics as a form of modernist mimesis, or an imitation of modernity, that belies a collection of political arrangements deeply rooted in colonial antecedents. The bodies of *mareros* and the things *mareros* did to bodies seemed to challenge the biopolitical mimesis of the Honduran state that mainly served to legitimize a postcolonial order of things: clientelism, institutional corruption, and profound socioeconomic inequities.

Compare, for example, the images of *mareros* in print media to the official state photograph of President Ricardo Maduro (2002–6). Maduro stands tall in the portrait. The president's neatly shaven face, the modern cut of his business suit (which is traversed diagonally by a blue-and-white sash bearing the national crest), and his immaculately styled salt-and-pepper hair convey the idea that this man is in his prime: mature, knowledgeable, and energetic. Maduro's body bears the distinctive features of regimentation that historians and anthropologists recognize as indicators of localized, bourgeois, bodily class formations (Stoler 1995; Foucault 1978). Claudio Lomnitz (2001) has also analyzed presidential imagery in Latin America as a form of hyper-reality that creates the illusion that a modern state and what it is supposed to comprise—its institutions, infrastructure, policies, and practices—exist in a national territory although the state only exists in the body of the president. Finally, Maduro's portrait urges one to consider the capacity of such imagery to instill and excite desires for specific kinds of embodiment (Farquhar 2002). Does the photograph incite the question, is this the person I want to be?

It is also noteworthy that *mareros* are not the first figures to be vilified as an external threat to normative society in the history of Honduras. In the region that is today's Central America and Mexico, the representation of a kind of otherness akin to that of the *mareros*—as a threat from without whose neutralization legitimizes colonial and postindependence state governments—can be dated to the sixteenth century. Colonial Spanish inquisitors in other parts of Central America and Mexico such as Diego de Landa, for example, viewed indigenous populations as terror-inspiring, savage, cannibalistic wielders of the devil's power, and the inquisitors justified colonial governance and violence as necessities for guiding these populations to (what the former deemed was) the proper path of Christianization (Coe 2012).

In the early sixteenth century, however, the area that became present-day Honduras did not have the higher indigenous population density or the centralized tributary sociopolitical systems of the adjacent Mesoamerican region. For historians Murdo MacLeod (1973) and André-Marcel D'Ans (1998), the relatively smaller, decentralized indigenous population of the Central American isthmus limited the settlement of Spaniards who sought to exploit the labor and tribute of non-Europeans. As a result, the Capitanía General de Guatemala and Central Mexico became the regional centers of Spanish colonial society, and Honduras turned into a peripheral area. For D'Ans the presence of a Spanish criollo class (people born in the colonies, descendants of Iberian parents who differentiated themselves from mestizo and indigenous populations through bodily cultivating practices meant to index their "Spanishness") was a necessary element to ensure the proper passage from colony to nation-state. In the absence of a substantial elite criollo class and a more populous indigenous labor force, D'Ans sees the cultural tone of colonial society in Honduras as having been established by the rudimentary miners (*mineros guirises*) who congregated around the mines of Tegucigalpa in the country's highlands.

His description of the *mineros guirises* paints an image of deficiency in terms of the behaviors that, from a foucauldian perspective, are indicative of social body formation among national bourgeoisies. D'Ans represents the *mineros guirises* as "impoverished, promiscuous, misceginating gamblers, with little concern for the future, who are content with the impromptu resolution of problems as a means of 'getting by'" (1998, 66; translation by author). In the absence of the criollos, D'Ans ethnocentrically sees Honduran colonial society as "sinking into ecological, economic and cultural decadence" (66; translation by author).

I interpret mainstream Honduran media representations of *mara* members as a similar trope of Honduran national historiography. The *marero*, like the *mineros guirises* and indigenous populations, is figured as a form of otherness that necessitates the existence of colonial forms of governance and postcolonial nation-states and legitimizes their tactics of paramilitary and state violence and *mano dura* crackdowns. The Honduran media renders the gang member's body as the implicit flipside of Maduro's presidential image. Its depiction is the affective, terror-evoking fulcrum upon which state power can be leveraged. It is also worth keeping in mind that the terror and voyeuristic pleasure gang stories arouse is not felt by a solely biological body; that body's response is also shaped through meaning-laden quotidian practices such as writing, reading, speaking, and listening. This body has a long history in the making, and its reactions of pleasure or terror are by no means rational, natural, or spontaneous but are situated in a rich matrix of colonial and postcolonial national culture.

Affective Ambiguity and the Limits of Terror and Desire

But how exactly did non *mara*-involved Hondurans experience representations of gang members? Did the images of *mareros* as an imminent threat to societal order result in a nationwide popular mobilization to bar them from the working-class neighborhoods they lived and operated in?

In August 2000 the Honduran daily journal *El Tiempo* published a story titled "La Flor del Campo Prisionera de Pandillas" (The Flower of the Countryside prisoner to gangs). The Flower of the Prairie was a suburban subdivision at the fringes of the Honduran capital of Tegucigalpa. The story included a photograph of several shirtless, tattooed *mareros*, posing with their hands contorted to demonstrate their gang signals. One of the *mareros*' foreheads was tattooed with the number 666. The figure simultaneously referenced Christian beliefs about the Antichrist and the Mara Dieciocho's emblematic number, which is the result of its sum. The image was taken indoors, and the article did not give the location where it was shot. Behind the menacing *mareros* was a white wall adorned only with a poster featuring the popular animated character Tweety Bird surrounded by stylized hearts.

As I read the paper at an Internet café in Tegucigalpa, the attendant,

Byron, peeked over my shoulder. Byron and I had become acquainted in the previous months, as I was a regular patron. He was in his early thirties and worked part-time at the café near the city's center. Byron lived in a precarious social situation: technically trained in computer maintenance after high school and with middle-class aspirations, he sought a job where he could apply his knowledge of computer hardware but, at the time of this ethnography, had not managed to secure such employment.

Seeing the photograph that accompanied the story I was reading, Byron spoke up: "Son mal*ditos*!" (They're god*damn*ed!) Byron's exclamation had a visceral glee that conveyed an ambiguous mixture of condemnation and admiration. The ambiguity of his comment was fueled by the *mareros'* proud display of their rebelliously tattooed bodies and by their having desecrated the unmarked skin, which had become a transnational sign of bourgeois health in the twentieth century (DeMello 2000). In their act of desecration, the *mareros* seemed to have undone the signs that marked their status as subordinated members of the *clase obrera*, and the hard-working and underpaid attendant longed, if for just one moment, to be a part of such an erasure of class and nation-forming desires.

The attendant's exclamation resonated with another story I had recently heard from a struggling writer and Internet café proprietor named Felix Castellanos. Felix was a college-educated, middle-class resident of Tegucigalpa whose café served multiple purposes. The café functioned as a place of socialization where steady patrons brought music and food to share with other regulars, workers, and owners; as an impromptu art gallery where Felix's acquaintances could display paintings; and as a place where foreign travelers could meet Tegucigalpans and ask for hotel and nightlife recommendations. Felix decided to befriend me because he thought I was a suitable candidate for a social circle he was carefully forming. The Internet café proprietor saw his identity as a writer threatened by what he thought to be the intellectual mediocrity of his fellow Tegucigalpan neighbors, confiding in me during a private conversation: "Ya no quiero hablar con gente pendeja!" (I am tired of speaking with dumbasses!) *Pendejos*, Felix explained, were people who lacked (what he considered) intellectual curiosity. They were the *mineros guirises* of his day. His careful selection of acquaintances—artists, academics, physicians, entrepreneurs—was a class-forming practice meant to differentiate him from his uncultivated compatriots.

One afternoon, after I told him of my interest in *maras*, Felix shared a story he had recently heard. In a hushed but excited tone, he told the following anecdote:

> A small community in the outskirts of San Pedro [Sula] was left disconnected from the outside world because the footbridge that spanned a nearby river was washed away during Mitch. The bridge was replaced with a metal structure by an international donor but was also washed downstream during another storm and remained unrepaired. A group of *mareros* who lived in the village decided to take matters into their own hands and reinstalled the bridge. From that day on, the *mareros* charged all who crossed for their use of the bridge. (unstructured interview 2000)

Felix narrated the story with a mixture of admiration and terror—admiration that the organization's efficacy at addressing public needs surpassed that of the Honduran government and international donors, as implied by the bridge's lack of repair, and fear that such efficacy is attributed to a group associated with unpredictability and violence in the Honduran and international mass media. The story took place in a remote and tropical location, the north coast, away from the temperate region of the Honduran capital. The combination of obscure actors with an exotic location had the cumulative effect of producing a sense of uncertainty that terrified and titillated its narrator and listener. Claudia Castañeda has interpreted rumors such as Felix's narrative as a "particular kind of story which can metaphorically describe fears and raise moral questions . . . the same story may also chronicle realities otherwise denied" (2002, 113). I would add that rumors are an affect-mobilizing part of the social realities lived by people; therefore, they should not be methodologically dismissed as falsities to be discriminated from ethnographic truth. Rumors are part of social reality because they are agential entities, mobilized by people, with the power to influence concrete actions and evoke affective reactions.

For the teller and listener of this narrative, whether the events Felix described actually happened is not as ethnographically relevant as the plausibility that such a sequence of events did occur, the effect that this likelihood has on their behavior, and the role that sensations such as terror and pleasure have in shaping this behavior. In Felix's case, a story of the *mareros*' repairing a storm-damaged footbridge in the context of a state's failure to address a disaster has the effect of making him hesitate to dismiss the *marero* as an imminent threat to Honduran society. While news and Internet media routinely represent the *marero* as lacking in ethics and reason, stories such as Felix's about adolescent gangs present these groups as informal organizations in which marginalized adolescents attempt to configure what Stephen Collier and Andrew Lakoff (2005) have called regimes of living—that is, a way to formulate contingent means for organizing, reasoning about, and living ethically in uncertain situations.

The affective ambiguity that both Felix and Byron experienced created turbulences in and inadvertent resistances to the way power operated through terror and desire in *mara* representations. The complexity of their affective responses created some possibilities for Hondurans to question hegemonic social orders and examine the unresolved tensions and contradictions of living in a former colony. To paraphrase Castañeda (2002, 113), rather than simply leading them to condemn *mareros*, some Hondurans used stories about *maras* as "a means to metaphorically describe fears and raise moral questions" not only about the behaviors of unruly adolescents but also about seemingly failing states and national societies.

Felix's story and the newspaper narrative about the *mareros*' killing one of their headmen in the San Pedro Sula prison for his lack of leadership are ambiguous because they simultaneously evoke contrasting sentiments of fear, admiration, and pleasure, but these emotions are not experienced in a sociocultural vacuum. The fear evoked by these *mara* stories emanates from the uncertainty that arises when a non-state-sanctioned organization usurps the state's monopoly on violence. The admiration, however, arises out of this same organization's (imagined or actual) ability to surpass the postcolonial state in its institutional effectiveness (repairing a fallen bridge), its transparency, and its accountability (killing a leader who has failed in his duties to the group).

Baleadas *at Midnight*

But how did *mara* hyperreality match up to my own ethnographic experiences?

Late at night on December 24, 2000, I accompanied Leah Getchell to Parque Las Mercedes, the park located in front of the Honduran Congress. The purpose of our visit was to bring *baleadas* (a popular Honduran fast food consisting of a flour tortilla stuffed with refried beans, sour cream, and cheese) to a group of homeless children and adolescents who identified themselves as La Mara Mercedes Loca (Crazy Mercedes Gang). The night was cool and foggy, and under a tree slept a group of nearly ten children and teenagers covered by cardboard boxes and a plastic tarp. Jose Luis, a seventeen-year-old member of the group, was awake and greeted us. We gave him the bags of food, and he nudged another child, waking and instructing him to distribute the food equally among the remaining Mercedes Loca members. As the youths passed the *baleadas* around, we sat with Jose Luis, talking about his travels throughout Central America, his stay in zone 5 of Guatemala City (an area known for its poverty and precarious social situation), his experiences smoking crack cocaine, and tattooing. When the youths finished the *baleadas*, our conversations fizzled, and the members of Mercedes Loca began to yawn and reach for their pieces of plastic and cardboard. As the teenagers fell asleep, the Covenant House worker and I parted.

Sharing the *baleadas*, executing a *mara* member for his leadership failure, and repairing damaged infrastructure that the local government had ignored, whether the product of a fantastic narrative or an ethnographic observation, suggest that *maras* offer their members a social framework that fulfills the expectations of efficacy, transparency, and equity that the collection of bureaucracies and institutions known as the Honduran state has yet to provide its citizens. But what of those ethnographic encounters with *mareros* that are not replete with the usual and comfortable anthropological tropes of subordinated subjects who speak willingly and who engage in structural-functional redistributive practices?

The following week I returned to Parque Las Mercedes during a weekday afternoon. The streets that had been quiet before were bustling with activity. Buses, cars, office workers in high heels, men in business suits leaving the National Congress, fast food vendors with impromptu stalls fashioned out of baskets and wooden crates, and lottery ticket vendors filled the previously serene urban landscape. The Mercedes Loca was there, but what had been a palatable ethnographic other no longer was. The pungent smell of shoe cobbler's glue emanated from small baby food jars that the adolescents held under their shirts. The narcotic effects of the glue were evident in the youths' eyes, which seemed distant and unable to focus, making it impossible to hold a conversation.

Two thin, scantly clad women in their early twenties with Mara Dieciocho tattoos on their shoulders were also present. The Mercedes Loca members introduced me to the Dieciocho girls, and the former's gestures suggested reverence for the members of such a renowned *mara*. Seeking to increase my credibility with the Dieciocho girls, the Mercedes Loca members insisted on revealing the tattoos I had acquired during my own adolescence, lifting the short sleeves of my shirt with their fingers made sticky by the cobbler's glue. The visiting *mareras* responded by making sexual advances. As she teasingly grabbed my upper arm with both hands, one said, "Dame un hijo" (Give me a child [a common Honduran phrase used playfully to acknowledge attraction]).

The fumes of the cobbler's glue made me nauseous, and this sensation became the bodily state through which I processed the vacant gazes of the Mercedes Loca members and the advances of the visiting *mareras*. In this moment, my own deeply embodied biopolitical affective disposition—being a middle-class, Catholic, Guatemalan Ladino— overwhelmed my anthropological sensibility. I became tense and felt a profound aversion to the sight and smell of young *mareros* sniffing glue. Leah sensed my affective reaction. "What's wrong?" she asked. "Nothing," I replied. "Don't panic," I thought to myself.

In the midst of this hectic scene, a police officer drove up on a motorcycle and called aside one of the young boys who was inhaling the glue's fumes. As I watched, I wondered if the child was about to run into trouble. The officer, dressed in gray polyester, skin-tight riding pants and a uniform dress shirt, extended her hand and asked the child to relinquish his jar of glue. The child willingly gave her the jar, and she reached into it with a piece of cardboard, procuring the viscous substance. Quietly I wondered if she was conducting a toxicology test. Before I could ask what was happening, the police officer leaned over and used the adhesive to repair the sole of her boot, which had detached from the leather upper.

The officer's request of a child's glue to repair her boot is reminiscent of Felix's story about broken bridges and toll-charging *mareros*. In this ethnographic vignette, the officer's asking for assistance from the *marero* results in an inadvertent subversion of the clear separation between state, civil society, and gang that is implied in many media representations of *mareros*. The child's providing help in maintaining a police officer's equipment brings the *marero* into a role of logistical support, suggesting that *mareros*, rather than being an external threat to Honduran civil society and state, are socially connected to non-*mara* members and state agents in a variety of ways.

Countering Affects: Mara *Sociality and Kinship*

One April evening in 2001, I found myself seated in the house of Doña Laura Martinez. Doña Laura's house had been washed away in the floods caused by Hurricane Mitch in Choluteca, and she had since relocated to Marcelino Champagnat. Following the disaster, her teenage son, Mario Jose, became involved with the Mara Dieciocho. Unlike the nearby resettlement community of Limón de la Cerca, where *mara* activities went nearly unchecked, some Marcelino residents organized informal vigilante groups aimed at curtailing the gang phenomenon. As we spoke about her son's gang involvement, Doña Laura seemed distraught. "Me lo van a matar" (They are going to kill him), she told me (fieldnotes 2001). She explained that adult men had come to her house earlier, searching for the boy, and had vowed to harm him if they caught him.

Mario Jose sat on a plastic chair, listening to our conversation. He was obviously displeased with his mother's affliction and protective condescension. He sat with his head slightly bowed, with his baseball cap hanging low over his forehead, preventing eye contact with either of us. His arms were tattooed, the right with the number 1, the left with the number 8. Together they made the number 18 when presented to an onlooker.

The threats made to Mario Jose were by no means idle. A few weeks later, a group of Marcelino residents accidentally killed a youth who was not involved with a gang and who was working for a reconstruction

NGO as a welder's assistant. The vigilante group had originally intended to kill a well-known gang member but mistakenly shot the other adolescent, who resembled their original target.

At a later date, I found myself sitting in the back of an urban bus going from Choluteca to Limón de la Cerca. Before leaving the city limits, five adolescents adorned with Mara Salvatrucha tattoos boarded the public transport. The seat next to me was vacant, and one of the teenagers took it, while the others remained standing in the aisle. The young *marero* took an interest in my bag, which contained nutritional survey materials, and asked about its contents. "What are you carrying in that bag?" he asked, motioning toward my backpack. "It's a scale to measure food portion sizes for a study on childhood nutrition I am working on," I said, as I opened the bag and showed him the contents.

With his curiosity sated, the teen shifted his attention from me to a commotion that had just started at the front of the bus. A law had been recently enacted that prohibited passengers to stand in urban buses to prevent overcrowding (*Diario el Heraldo* 2000b). The law was passed after a highly publicized traffic accident in Tegucigalpa, in which a pickup truck carrying the better part of a youth league *fútbol* team swerved at high speed and catapulted the majority of the young men traveling in its bed, causing the deaths of four of them. Although most Hondurans I encountered recognized the soccer players' deaths as a tragedy, many also perceived the government public safety office's response—disallowing passengers to ride in truck beds or to stand in bus aisles—as too severe a rule in a country where people relied heavily on public and shared transportation.

The commotion that drew the *marero*'s attention involved a seemingly drunken traffic officer dressed in a tattered uniform who had boarded the bus and ordered several women returning from the central market with tubs full of tortillas to exit the vehicle because they were standing in the aisle. When the women refused, the officer drew his revolver from its holster and aimed it at them, drawing loud jeers from the bus's passengers. The *marero* joined his friends as they pushed their way to the front of the bus, defiantly raising their shirts and revealing several knives and sharpened screwdrivers tucked into their pants. While I had previously seen the other passengers as a mixture of market vendors, tattooed adolescents, construction workers, and farm day laborers, they quickly coalesced into a constituency focused on expelling the traffic officer from the bus. With pushes and shoves, the passengers removed the policeman. As the bus pulled away and continued along its route, the *mareros* and market women rushed to the bus's windows, yelling, "Perro! Ba*ss*ura!" (Dog! Trash!) (fieldnotes 2001).

Doña Laura's words of concern and the expulsion of the traffic officer from a bus in Choluteca urge me to think of what the news stories about *mareros* and the cultural studies–type analyses of the power of representations routinely omitted: the emotion-laden social and kin relations that tied gang members to family and other residents of the neighborhoods they lived and operated in, and how these relationships posed a countertendency to the way power operated through terror in *mara* media stories and urban myths. Doña Laura was not alone in her attachment to *mara*-involved relatives. For every one of the estimated five hundred *mareros* operating in Cholutca at the time of my ethnographic work there was an extended network of friends and family who cared about their well-being.

While this observation may seem a truism at first, it is an important and often forgotten dimension of the street-gang phenomenon. After conducting a public presentation about my research at the university where I work, for example, a colleague who also works in Central America once commented, "Even from an emic perspective, people in poor neighborhoods want the *maras* gone." Such a statement assumes that *mara* members themselves are not legitimate members of subaltern neighborhoods in Central America, that any anthropological analysis that suggests anything but the swift eradication of *maras* stands on dubious ethical ground, and that "poor Central Americans" agree on how to deal with the issue.

Terror-evoking depictions of *maras* in the mass media routinely present the gang member as the location of intervention and rehabilitation, sustaining the necessity of the U.S.-Honduras relationships and the national development and class-forming practices that made gang identities possible. Maintaining the *marero* as the sole locality of state intervention (the annihilation or incarceration of *mareros* by paramilitary organizations and penal institutions) in Central America limits an engagement of this problem to a symptomatic level and sustains the societal processes that dialectically produce the phenomenon. Practitioners, whether applied anthropologists, concerned citizens, government officials, or journalists, must strive toward a redefinition of the *mara* problem that recognizes the quotidian ontological violences of class and nation formation that give shape to everyday forms of insurrection such as adolescent gangs. Only by identifying, reflecting on, and modifying such practices can the *mara* phenomenon be effectively addressed.

5. Ecologies of Affect and Affective Regimes

THE NEOLIBERAL RECONSTRUCTION
OF NEW ORLEANS

Can a plan devised by disaster recovery experts such as architects and urban planners for the reconstruction of a city make a person upset? Can its words and images elicit bodily effects like the tightening of muscles, the quickening of pulses, and the shortening of breaths on that place's residents? Many New Orleanians from the city's central neighborhoods who participated in its three recovery-planning processes after Hurricane Katrina would answer these questions with an emphatic *yes*. If this is the case, then what is an anthropologist to make of such manifestations of emotion? I would not have considered these questions prior to embarking on the ethnographic project that concerns this chapter—the ethnography of recovery planning in post-Katrina New Orleans.

In the pages that follow, I make the case that the manifestation of emotions in the context of post-Katrina recovery planning occurred at the point where New Orleanians' ecologies of affect met with what I call the affective regimes of state-sanctioned recovery planning. By using the term *ecologies of affect*, I intend to communicate the connection between the ethnicized, racialized, and class-based relations involved in making the city's built environment and social order and people's embodied sensibilities. Over the course of the last three hundred years, many New Orleanians have tried to realize a racially structured society by giving spatial form to their ideas about difference; that is, they have built neighborhoods where racialized class distinctions are made and sustained through everyday practices of space and body making (Breunlin and Regis 2006; Campanella 2006; Dawdy 2008; Hirsch and Logsdon 1992; Johnson 1992; Lipsitz 2006; Regis 1999).

The spaces and body politic of New Orleans, I would learn over the course of this work, constituted an ecology of affect where people came to experience varying attachments and preferences contingent on their life experiences and differential treatment. I choose the word *ecology* because, building on Tim Ingold's (2000) theorization of the term, it conveys the idea that people emerge as particular kinds of persons with values and distinct ways of experiencing emotions in the midst of multidimensional, open-ended, and co-constitutive meaning-laden relationships with other people and their surrounding material environment.

New Orleanians who had experienced imposed limitations on their social and spatial mobility manifested emotions such as anger and frustration when professional planners presented them with recovery plans that threatened to disrupt the practices they used to subvert the city's landscape of inequity. Through these practices subaltern New Orleanians made meaningful lives in an often-hostile society. Such recovery plans were upsetting because they upheld logics of capital investment and stranger sociality—the idea that cities are spaces inhabited by strangers who interact with each other as strangers—as uncontestable facts of disaster recovery planning.

The recovery plans that concern this chapter conceptualized the reconstruction of New Orleans as the creation of relationships between architectural structures conducive to the investment and circulation of capital and not the reinstatement of the city's pre-Katrina population, especially African Americans of modest means.. These plans also articulated their creators' ideas about what kinds of spaces, aesthetics, and modalities of sociality people should be endeared to. The spaces and styles featured in the plans were inspired by what Frederic Jameson (1991) calls the cultural logic of neoliberal capitalism. In this cultural logic, what matters are the colors, textures, and surfaces of spaces whose forms break with local convention and obliterate a site's history all in the name of an appearance that indexes a global cosmopolitan style (think of the interior of a Starbucks café but rendered at the scale of a city neighborhood). In the case of New Orleans, however, the aesthetics and spaces of recovery plans were not so much an imposition of a global cosmopolitan style as they were an adaptation of this style to architectural forms and historic landmarks thought to be iconic of the city. I use the term *affective regime* when referring to the expert planners' visions of urban recovery because it conveys the notion of governmental agents (i.e., expert planners hired by the New Orleans City Planning Commission and city council) imposing a system or planned way of feeling and emoting in relation to specific kinds of urban environments, a regime that can be contrasted with New Orleanians' ecologies of affect.

I also find it worth noting that the affective regimes of state-sanctioned recovery plans in New Orleans had a distinctly neoliberal aspect. Since the mid-2000s, a number of scholars and social commentators have called attention to the ways some policymakers, NGOs, members of the non-profit sector, and developers use disasters as opportune moments either to impose capitalist policies (Gunewardena and Schuller 2008; Klein 2007) or to reinvent capitalism (Rozario 2007). New Orleans was no exception (Adams 2013; Johnson 2011). In Katrina's aftermath, the city witnessed experiments with public-private partnerships where neoliberalism manifested in various ways: channeling state funds to for-profit firms specializing in the delivery of aid services as a mechanism of disaster response (Adams 2013), deregulating labor markets to facilitate the exploitation of U.S.-born and immigrant reconstruction workers (Button and Oliver-Smith 2008; Fussell 2007), and privatizing spaces and institutions that were previously used to provide public services such as education and housing (Barrios 2011). To this list, we may add the aesthetic refashioning of the city proposed in the recovery plans that was to serve as the affective economic engine of a neoliberal New Orleans.

The ethnographic vignettes included in this chapter feature instances when planning experts insisted over the course of participatory meetings that neighborhoods were best conceived of as sites of capital investment and that such a way of thinking was an uncontestable fact of disaster recovery planning. I consider such recommendations to be a neoliberal injunction to subject all aspects of human life to rubrics of capitalist cost-benefit analysis. At the same time, the vignettes show that many New Orleanians found such conceptualizations of neighborhood space to be potentially destructive of their ways of life, socialities, and space-making practices. Perhaps most important, neighborhood residents passionately voiced their counter-discourses in ways that conveyed a sense of their emotional involvement in the city's recovery.

The ethnography of recovery planning in post-Katrina New Orleans, then, shows that emotions and their manifestation have much to tell us about the relationships between the ways people inhabit their bodies (with their varying dispositions, tastes, loves, and demeanors), the ways people interact with the material world to produce built and natural environments, and the ways people devise meaningful and emotion-laden connections with one another—that is, what they care about and how they care about it. For these reasons I also make the argument that using emotions as a point of ethnographic engagement in post-disaster recovery stands to inform expert planners in devising what are often called culturally sensitive and location-relevant reconstruction policies.

My concern with the relationship between affect, space, and sociality is important because disaster response and recovery experts often see the manifestation of emotional states such as anger as behaviors unbecoming proper participation in the public sphere. In the ethnographic examples that follow, professional urban planners often perceived angry or distraught people as a form of disorder that was out of place in what they deemed proper public engagement. But what if the exclusion of such emotions and emotional people came at the cost of actually understanding the meanings and uses of space in a disaster-affected site and addressing the social dimensions of disaster vulnerability?

The Ethnography of Emotions in Recovery Planning

My ethnographic approach to disaster recovery in New Orleans developed from the experiences I had as a novice anthropologist in post-Mitch Honduras. In Choluteca I learned that the policies and practices of aid organizations and government agencies can articulate implicit assumptions about the nature of people, communities, and social well-being. I also learned that these assumptions do not always map neatly onto people, their relationships with others, and the ways they structure and experience space in their communities. Most important, I came to recognize that the rigid application of these assumptions perpetuated the undesired impacts of disasters.

As I thought about putting together a research project in a new locality, I tried to be mindful of the lessons I learned in Central America. The following questions immediately came to mind: Given the reconstruction complications Choluteca ran into, what would happen in New Orleans, a city with considerably greater socioeconomic and cultural complexity? Who would make decisions about the city's reconstruction and in what contexts? How would these decisions transform into practice and by what actors and to what effects? What ideas about well-being and personhood would be implicit in these decisions? Furthermore, what spaces were open to an anthropologist where I could observe and document the disaster recovery process?

In the summer of 2006, I continued to reflect on these concerns as I drove from Carbondale, Illinois, where I had just been hired as an assistant professor, to the flooded home of my parents—Teresa and Everardo—in Gentilly, a New Orleans neighborhood originally constructed during the first half of the twentieth century. During the initial round of fieldwork, I shared a trailer from the Federal Emergency Management Agency (FEMA) that was installed on the front lawn of Teresa and Everardo's home and devoted my time to reading through newspapers and local newsletters and to speaking with New Orleanians involved in community development. I was always looking for ideas on where to begin.

Two sites that stood out as worthwhile points of ethnographic engagement were the weekly meetings of the Neighborhoods Planning Network and the public meetings of the Neighborhoods Rebuilding Plan. Civically engaged residents who wanted to keep their fellow New Orleanians informed about the recovery process created the former group as a recovery-planning information clearinghouse, while the city council organized the latter as an ongoing recovery-planning process that featured public meetings in affected neighborhoods. Methodologically, I intended to use these meetings to see what ideas urban planners were entertaining on how to reconstruct the city. I also wanted to learn about the ways residents of various socioeconomic backgrounds responded to these ideas, and I hoped the discussions held during these meetings would help me learn who the actors involved in recovery planning were and what interests they had in the reconstruction.

The Neighborhoods Planning Network and Neighborhoods Rebuilding Plan meetings certainly had significant limitations as research sites. As I would soon discover, recovery planning in New Orleans was a high-stakes game with a number of prominent developers and politicians positioning themselves to influence and benefit from the city's reconstruction. Some of these actors preferred to remain out of public view, so closed-door meetings characterized the process just as much as the public gatherings that were accessible to me as an ethnographer. In fact, many residents spent the better part of their time trying to figure out the agendas behind the different recovery-planning initiatives and were always on guard that the official discourses iterated during public meetings never fully revealed the political and economic interests at play in the reconstruction.

The participants at these meetings also did not represent the entire cross-section of New Orleans society. The meetings and their professionalized structure of participation alienated some residents who were accustomed to other forms of community organizing that are often described as informal but are key to getting things done in some parts of the city. Remaining mindful of these limitations, I used the meetings as a point of ethnographic entry and not as the ethnography's permanent site. Thus, my ethnographic method became nomadic, moving through spaces and social relations beyond the official recovery-planning meetings and into the neighborhoods and households of those participants who were willing to share their lives and friendships with me.

Picking the recovery-planning process as my ethnographic point of departure proved difficult for some of my colleagues to comprehend. A friend and fellow anthropologist with whom I shared early drafts of this chapter once questioned the merits of my methodological choice, asking: "I thought ethnographies were something you did with communities? This is a recovery-planning process, and a plan is just a plan anyway. It will just sit on a shelf." While my choice of recovery planning emerged out of the sociopolitical and methodological challenges of doing anthropology in post-Katrina New Orleans, my choice was also informed by the science and technology studies literature where anthropologists, philosophers, sociologists, and historians are accustomed to viewing scientific experiments and laboratories as objects of inquiry. These sites, rather than being spaces evacuated of cultural values and of little interest to an anthropologist, were what Donna Haraway (1997) called spaces of cultural implosion. In these locales, meaning and materiality are fused and reconfigured into new forms that drive processes of cultural change. Granted, recovery-planning meetings are not laboratories. They are spaces filled with different actors, practices, politics, and discourses, but the science and technology studies literature was helpful when pondering the methodological merits of recovery-planning meetings.

Rather than being separated and abstracted from a broader society, recovery-planning activities turned out to be key sites where the cultural politics of urban space in New Orleans regularly played out. At these meetings, I was able to observe contestations over the imaginings of urban space among residents and professional planners. In these activities, if I paid close attention, I could also meet people who attempted to voice the sentiments and priorities of many of their neighbors, who either could not or would not participate in official planning processes.

My work in New Orleans, however, presented a number of different methodological challenges when compared to post-Mitch Choluteca that went beyond ethnographic site selection. Southern Honduras was a site that fit more comfortably within customary notions of ethnographic research. Choluteca was a compact city that was easily manageable in terms of identifying people I wanted to get to know and speak with. The resettlement sites were easy to delimit geographically, and somewhat effortlessly I established rapport with interview respondents and ethnographic interlocutors. This methodological ease was due, in part, to how Cholutecans perceived me and to the ways I inevitably benefited from both the region's colonial history and the hegemonic relationship of the United States with Honduras. Cholutecans often described me as a gringo or a *chele* (a colloquialism meant to signify "whiteness") even though I often tried to explain my liminal status in U.S. society as a transnational migrant from Latin America. My assigned status as a gringo in Honduras was a complex one. Sometimes my privileged connection with the United States justifiably earned me the contempt of some Cholutecans, but most associated me with U.S. cultural influence, making me a desirable person to speak with. Their perception of me, combined with their customary hospitality, meant southern Honduras was a relatively uncomplicated place to conduct ethnography. Of the 232 homes whose doors I knocked on, only two heads of household refused to speak with me during my thirteen-month stay.

New Orleans, in contrast, presented a number of different complications. Because of the city's history of race and class relations, the neighborhoods where I chose to work were contested spaces, and my movement through them as an ethnographer could at times elicit complex sentiments among subaltern New Orleanians. Upon learning I was an anthropologist, for example, Mr. Lionel Jones, a prominent resident of Tremé (one of the city's central historic areas) who had a lengthy history of participating in the neighborhood's cultural life (second-line parades, Mardi Gras), told me: "I will never let you interview me. You can ask me to have a conversation, and you can come to my house and we can talk like people, but you can't interview me" (fieldnotes 2007). What Mr. Jones meant to communicate to me was that in doing the ethnography of recovery planning, I had to carefully reconsider my methods and expectations. His initial refusal to be interviewed, he would later inform me, was not necessarily because he wanted to be kept out of my research but because he wanted to exercise his agency on how I treated him as an interlocutor. Consequently, I was going to have to tread lightly in a place where I was an outsider and where everyday battles over the use of and right to urban space had been playing out since the city's foundation in the eighteenth century. Rather than relying on a techno-scientific sampling method to conduct hard and fast ethnography, I was going to have to polish my interpersonal skills and learn the nuances of relationship building in New Orleans. Instead of relying on a large sample of interview respondents, I would have to cultivate long-term relationships with a select number of residents and use our shared experiences in neighborhood organizing and quotidian activities as a vantage-point for trying to understand the city's neighborhoods, what they meant to their residents, and how these meanings manifested affectively.

Two things became apparent during the early phases of this project—the importance of neighborhood identity in New Orleans and the significance of the central neighborhood of Tremé. When I began this ethnographic endeavor, I intended to organize my work at the scale of the city, but I quickly learned from attending the planning meetings that I had to understand the subtleties of sub-city-level identifications, spaces, and politics. What also became apparent was that among the seventy-three neighborhoods officially recognized by the New Orleans City Planning Commission, the identifications, spaces, and politics of Planning District 4—and Tremé in particular—often provoked the most controversy and emotion-laden debate when discussed during recovery-planning meetings.

Planning District 4 is directly adjacent to the city's tourism and commercial centers of the Central Business District and the French Quarter, and it extends in a northwesterly direction toward City Park (fig. 10). The district includes some of the city's earliest expansions beyond the colonial French Quarter, and these neighborhoods, Tremé included (fig. 11), date to the late eighteenth and early nineteenth centuries. The area features historic architecture such as shotgun homes (their rooms extend from a main hallway that runs the length of the house from front to back) and historic landmarks like St. Augustine Catholic Church, which is credited with serving the first racially diverse Catholic congregation in the United States. Before Katrina, 77 percent of the district's residents rented the homes they lived in, and the median household income for the area was $17,830 (GNOCDC 2007). District 4 was also home to four of the major public housing projects in the city—Iberville, Lafitte, St. Bernard, and C. J. Peete—some of which were the first of such structures built in the country following the Second World War. I did not initially intend to focus my ethnographic work on Planning District 4 (and Tremé in particular), but its residents and their critiques of recovery plans quickly brought the site to my attention.

Space and Affect

The lessons I learned from my work in post-Katrina New Orleans went beyond methodological ones. Prior to beginning this project, interrogating space as not just a neutral and objective backdrop where human action takes place (what some might call etic space) but, instead, as something that is dialectically produced through people's practices in simultaneously epistemic, material, symbolic, and affective ways had not captured my interest. My ethnographic experiences in New Orleans and the dialogues I had with colleagues about them urged me to take a closer look at the observations of critical anthropologists and geographers who had, for quite some time, called for the recognition of space as a social product. Particularly helpful in this regard was Henri Lefebvre's (1992, 1996) claim that space is socially produced and that its production has three interrelated moments: the *lived*, or the actual built environment;

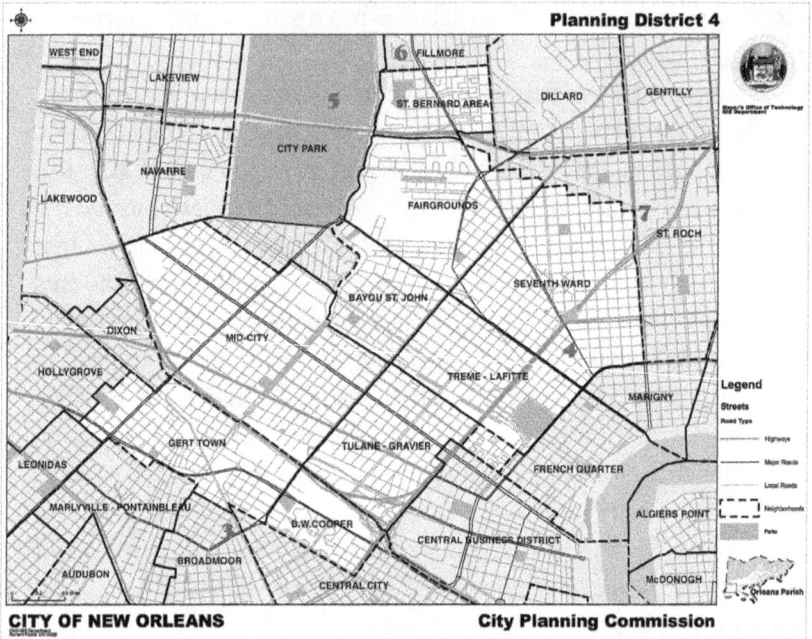

FIG. 10. New Orleans Planning District 4. Courtesy of New Orleans City Planning Commission.

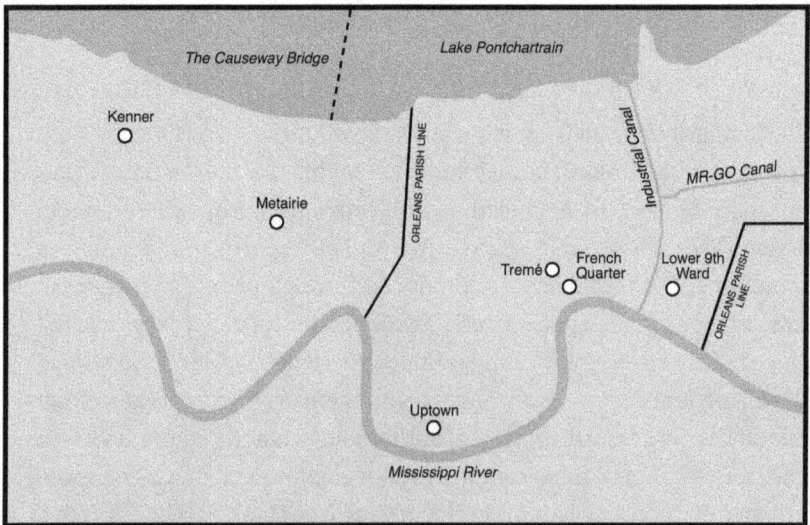

FIG. 11. Location of Tremé in relation to other New Orleans neighborhoods and suburbs. Courtesy of author.

the *conceived*, or the visions, plans, and designs—often on the part of politically and economically influential groups—of space entertained in a particular social context; and the *perceived*, or the memories, rituals, and meanings people have of, perform in, or develop in a particular space, or what some social scientists would distinguish as "place."

Lefebvre's triad proved a useful point of analytical departure when considering the power differentials of whose voices and ideas were actually incorporated into recovery plans and whose voices were silenced, as well as understanding how the built environment of New Orleans came to be and the ways New Orleanians of various socioeconomic backgrounds appropriated and felt about this space. Proving equally useful was Setha Low's (2000, 2003, 2011) theorization of space as the result of the *production* of space, or the way the built environment is created and the way issues of power and hegemony play out in this process; the *construction* of space, or the meanings people come to attribute to the built environment; and the *embodiment* of space, or the ways people come to be in their bodies as a result of life experiences in socially constructed and socially produced spaces.

It is noteworthy that Low, like Lefebvre, makes an important differentiation between the material dimension of space (lived space and socially produced space) and the affectively experienced, intersubjective dimension of that space (perceived space and socially constructed space) that is also often recognized as "place" among anthropologists (Low and Lawrence-Zúñiga 2003). Place, then, is the unique and varying way in which people experience space and come to associate specific memories and sentiments with it. Finally, Rachel Breunlin's (Breunlin and Regis 2006) and Helen Regis's (1999) spatial analysis of ritual practices such as second-line parades and race relations in New Orleans also helped guide my thinking about the connections between space, practice, and affect in recovery planning.

A critical reader may wonder how space is socially produced and constructed and might ask, don't Cartesian maps, Google earth, and the Global Positioning System (GPS) show us space as it really is? Does the space illustrated on the screens of our navigation computers not exist regardless of human action? Both Marxist and Foucauldian scholars have pondered this question for some time. Claude Raffestin (2007), for example, insists that the aforementioned are representations of space, and, as such, they are always mediated by some epistemic lens, or some way of seeing the world and interpreting it through culturally situated categories and values that are by no means objective. For example, rather than being an objective means of seeing the world, GPS is the product of a North American military-industrial complex, and the system's initial purpose was to guide military munitions to their targets. Do GPS systems work? Absolutely. But consider also Erin Manning's (2012) observations about Australian Aboriginal sand paintings, which are often relegated to the status of culture and differentiated from technology (which GPS systems are readily associated with). Manning argues that Australian Aboriginal sand paintings are filled with information about the environments that surround their painters, such as the location of water sources and sacred sites, and are therefore as technological and effective as a GPS system.

The key idea here is that both GPS-derived maps and sand paintings are representational dimensions of socially produced space. As such, they are culturally contingent, but this cultural contingency does not deprive them of practical capacities for the people who use them. In fact, their cultural "situatedness" defines their usefulness. Nevertheless, as culturally specific ways of representing space, they are also tied to deeper and broader cultural histories of seeing, engaging, representing, and making the material world. The important point is that space, first and foremost, is a human concept and space making is a human practice. The idea that space precedes human action—be it intellectual, architectural, or environmental—and can be accessed at an objective "etic" level misses this important observation.

What follows is not a tedious triaging of ethnographic information into Lefebvre's three moments and Low's three processes. Instead, it is an ethnographic narrative that is informed by the fundamental insights of these scholars' work: Space does not precede social action but is made through people's practices, the built environment is always the materialization of complex processes that involve people's culturally situated imaginations and desires, and people who live in such socially produced space always subvert hegemonic desires by devising their own meaning and personhood-making practices in a given locality. Furthermore, my work in New Orleans pushed me to explore the spatial dimensions of the ecology of affect—that is, the ways the sensing body comes into being in a world whose spaces are made, remade, and contested at the nexus of human-material relationships.

The Historicity and Ecology of Affect in New Orleans

During its three-hundred-year history, New Orleans has been a place where people make racialized differences and inequities by socially, spatially, and materially structuring the city's urban landscape. Throughout the city's historical development, many residents have denoted racialized identities by using location (where they live) and practice (what they do where they live)—human actions that eventually become embodied (Bourdieu 1977)—and these factors have influenced the ways people experience affect. Allow me to clarify that I am not suggesting all New Orleanians engaged in the same practices of space and difference production that created the city's landscape of inequality. The last three centuries feature multiple instances when residents subverted the city's racialized body politic through their social relations, intimacies, places of residence, and space-making practices. Nevertheless, the city's history also reveals salient hegemonic tendencies that had memorable and tangible impacts on the lives of many North Americans.

When the city was first founded in the eighteenth century, ownership of property in the area that is today's French Quarter was legally limited to people who could claim a full French genealogy (Johnson 1992). Under French colonial law, African slaves could earn their way out of bondage, and by the second half of the eighteenth century, the city counted a growing population of freed African and African American residents. In addition, unions (as well as sexual violence) between French, African, and Native American populations gave rise to a new group of francophone people who did not fit existing categories of racialized differentiation. In the nineteenth century, these people came to self-identify as Creole (Dawdy 2008).

As described in chapter 4, the word *Creole* as a means of identification originated in the Spanish colony of New Spain; people used it to refer to American-born descendants of Iberian settlers (Cañizares-Esguerra 2002; Stoler 1995). Members of other colonial powers, including the Dutch and French, eventually borrowed and adapted the concept (Dawdy

2008; Hirsch and Logsdon 1992; Stoler 1995). In New Orleans, people initially used the term *Creole* in a similar fashion but to denote the American-born descendants of French settlers in Louisiana. Eventually, as in other parts of the Americas, people began using the term to categorize the emerging group of people who counted Europeans, Native Americans, and Africans among their ancestors (Gordillo 2014). In both usages, the idea that cultural practice and its embodiment are means of denoting racialized identity remained a central value. The term *Creole*, then, was first and foremost used to denote the novel form of embodiment attained through the cultivation of the body in a multicultural context where behaviors thought to index "Frenchness" (e.g., language, etiquette, dress) were considered superior to the cultural practices of Native Americans and Africans.

In the late eighteenth century, freed Africans, African Americans, Creoles, and recent European arrivals who were viewed as cultural others by established French settlers began to acquire lands in areas extending away from the French Quarter (Campanella 2006). The area that is today's neighborhood of Tremé was a nonfunctioning plantation belonging to Claude Tremé, a French settler who came into possession of the land through marriage. Tremé partitioned the plantation's land and offered it for sale to all those colonial others who had limited options for property ownership in the city (Toledano et al. 1980).

Before his death in 2011, Collins "Coach" Lewis was known for his participation in Tremé's cultural life as a Mardi Gras Indian. He and his fellow pedestrian parade participants organized in tribes and wore costumes inspired by stylized Great Plains Native American culture (fig. 12). After Katrina he spoke of Tremé's early years by alluding to the racialized spatial constraints of life in New Orleans:

> This area was diversified. You had a lot of people, black and white, come out of this area. When the boat landed, you came to Tremé cause that's where you were accepted. Everyone got together here and the music and food and culture come out of that. (unstructured interview 2010)

In the decades immediately following its founding, Tremé was a diverse neighborhood, but this diversity began to shift into racial polar-

FIG. 12. Mardi Gras Indian Parade. Courtesy of author.

ization during the nineteenth century. The city's Americanization after the Louisiana Purchase in 1803 led to the arrival of new settlers from former English colonial domains who conceptualized racial difference as a black-white binary. This way of thinking about—and making—racial difference contrasted with French colonial notions of race; the French emphasized the role of language and those cultural practices they considered indicative of "Frenchness" for defining in-group membership. While Creoles could never make unqualified claims to being French, they could become "more French" through cultural behavior. The black/white binary of Anglo America, in contrast, set a much more rigid way of categorizing New Orleanians and thus threatened the social status of many Creoles. To add another layer of complexity, the city's population doubled in 1810 after the arrival of refuges from Saint-Domingue, and this group observed notions of racialized difference similar to those of francophone New Orleanians (Campanella 2006).

After the Civil War, New Orleans also witnessed the migration of emancipated African Americans into its urban areas. The black-white binary limited the possibility of Creoles to socially differentiate themselves from recently emancipated slaves. Consequently, many Creoles established new neighborhoods such as the Seventh Ward to the north and northwest of the French Quarter in an effort to distinguish themselves from working-class African Americans (Campanella 2006). (The term *ward* is used to describe city voting districts, which were first delimited in 1809.)

The meanings of nineteenth-century neighborhood spaces have endured in New Orleans, and they still surfaced over the course of casual conversations after Katrina. Anthony Robinson, my parents' neighbor, for example, grew up in the Sixth Ward, where Tremé is located, and his wife grew up in the Seventh Ward. During a summer day in 2007, I caught up with him as he was returning from a fishing outing with his cousin. Anthony boasted about his fishing prowess, mentioning that this ability had earned him his father-in-law's approval. His wife's father, Anthony explained, initially viewed him with skepticism because of his neighborhood of origin. With a little bit of glee in his voice, Anthony recalled, "He thought I was too slick for his Seventh Ward daughter!" (fieldnotes 2007). But he claimed he overcame this initial suspicion with gifts of fish procured from the city's bayous and canals.

During the nineteenth-century era of state-sanctioned segregation, working-class African Americans also created parading, ritual, and masking practices whose diasporic histories have West African origins (Lipsitz 1988; Regis 1999). Under Jim Crow laws, African Americans found themselves cut off from access to public services and resources. Consequently, they created organizations called social aid and pleasure clubs that would provide their members with assistance during key life events: weddings, funerals, and times of hardship (Breunlin and Lewis 2009; Lipsitz 1988). One of the ways these organizations financed themselves was through leading weekly pedestrian parades that traversed various parts of the city and collecting donations. The parades featured a brass band and a group of uniformed dancers, which were referred to as the "first line." Additionally, a group of spectators formed and walked behind the parade as a "second line," for which those parades are named (fig. 13).

Second-line parades subverted the city's spatialized body politic by providing a mechanism for its participants to move through areas where the residence and presence of African Americans were formally and

FIG. 13. Nine Times Social Aid and Pleasure Club second-line parade. Courtesy of author.

informally limited (Breunlin and Regis 2006; Lipsitz 1988; Regis 1999). In addition to second lines, working-class African Americans devised other parading practices such as Mardi Gras Indian Super Sunday. As noted previously, Mardi Gras Indians are parade participants who, with the assistance of kin and friendship networks, craft elaborate beaded costumes based on a stylized Great Plains Native American aesthetic and are organized in ranked groupings known as Mardi Gras Indian tribes (Breunlin and Lewis 2009; Ehrenreich 2004; Lipsitz 1988). New Orleans grassroots historian Ronald W. Lewis explains the origin of this particular parading practice by making explicit connections between the use of space in the making and sustaining of racialized difference and the experience of affect:

Coming out of slavery, being African American wasn't socially accept-able. By masking like Native Americans, it created an identity of strength. Native Americans under all pressure and duress, would not concede. These people were almost driven into extinction, and the same kind of *feeling* came out of slavery. "You're not going to give us a *place* here in society, we'll create our own." In masking, they paid respect and homage to the Native American by using their identity and making a social statement that, despite the odds, they're not going to stop. (Breunlin and Lewis 2009, 65; emphasis added)

The end of state-sanctioned segregation in the early 1960s marked another pivotal moment in the city's history of spatialized differentiation. The desegregation of schools and public housing facilities instigated a racially motivated suburban flight of those residents who self-identified as white. Between 1960 and 2000, the city lost more than two hundred thousand people (Campanella 2006). At the same time, new suburbs such as Metairie and Chalmette sprouted to the west and east of Orleans Parish lines, and their populations grew rapidly as their inhabitants attempted to spatially distance themselves from African American New Orleanians.

Quotidian practices of spatial separation and de facto segregation were also complemented by unofficial policies of suburban police departments. In the 1990s the Jefferson Parish Sheriff's Office became nationally renowned for its unabashed racial-profiling activities, which were meant to limit the movement of African Americans through the suburbs. Sheriff Harry Lee, the department's top authority, went on record saying, "Why should I waste time in the white community" (Burnett 2006). He also was quoted as saying, "We know where the problem areas are. If we see some black guys on the corner milling around, we would confront them" (Nossiter 2007). Harry Lee himself demonstrated the complex cultural dimensions of racialized identity in southeastern Louisiana. A descendant of Chinese immigrants and a self-identified Asian American, Lee also engaged in practices meant to index southern working-class white masculinity. An avid gun collector, hunter, and binge drinker, he proclaimed himself a "Chinese Cajun Cowboy" (Burnett 2006).

Returning to that summer day in 2007 when Anthony Robinson had come back from his fishing outing, my experiences in New Orleans instructed me on how those residents whose social mobility was constrained by the spatialization of New Orleans's body politic came to sense emotions like indignation. Without having to explicitly mention racism, Anthony, while cleaning fish on a cooler outside his house, communicated his feelings about the treatment of African Americans in New Orleans's suburbs: "I don't go shop there [Metairie]. Harry Lee, Sheriff [Charles] Foti [of Orleans Parish], and who is that guy who owned all those restaurants? Al Copeland [the founder of the fried chicken restaurant chain Popeyes]. They're all the same" (fieldnotes 2007). By making the association between Harry Lee and Sheriff Foti, Anthony referenced the former's reputation as an upholder of racial-profiling practices and associated it with the latter. Furthermore, by including Al Copeland in this list of objectionable prominent citizens of Jefferson and Orleans Parishes, Anthony went on to denounce the suburban business elite as complicit in the segregationist practices of the greater New Orleans area.

The suburban flight of New Orleanians who self-identified as white was accompanied by what Antoinette Jackson has called the pre-Katrina diaspora of middle-class African Americans (Jackson 2011). This diaspora consisted of people who had left Louisiana in search of employment opportunities in professional labor markets that were not as badly plagued by racism and people who had moved to other suburbs such as New Orleans East (Sorant, Whelan, and Young 1984). The post-desegregation suburban flight resulted in the proliferation of blighted properties, the racial polarization of many New Orleans neighborhoods, and the isolation of working-class African Americans from important services such as well-funded public schools and employment.

In the 1960s a number of urban renewal projects further maligned Tremé's commercial areas. One of these projects featured the construction of an elevated highway over North Claiborne Avenue that dealt a heavy blow to many of the adjacent businesses (CUPA 1995; Lacho and Fox 2001). Despite these impositions on the part of city planners, Tremé's residents reappropriated the spaces underneath the elevated highway, using them as locations to conduct second-line parades, Super Sunday celebrations (Breunlin and Lewis 2009), and daily socialization. In this ecology of affect, sociopolitically marginalized New Orleanians gave new meanings and purposes to urban spaces and developed unique emotional attachments to them.

In the 1970s city planers also authorized the construction of Louis Armstrong Park, a project that displaced more than 160 families. The undertaking included the construction of a concrete fence around the park that limited the mobility of Tremé residents through its grounds. After Katrina those families who had lived in the area for the previous four decades still remembered the enterprise as an arbitrary imposition by the city government that had an ethnocidal impact on their neighborhood because it displaced families that were considered a part of the area's social fabric.

In the aftermath of Katrina, residents who had lived in Tremé since the late 1960s emphasized the role of social relationships among neighbors, the specific uses of space, and the importance of particular places when defining what the neighborhood *was* and what they cared about in recovery planning. More than a geographically delimited space or a collection of historic architectural structures, for them, the neighborhood comprised uniquely significant places such as bars and lounges (fig. 14) that served as spaces of daily socialization, and it was the social relationships that people built and enjoyed in these spaces that came to define Tremé's character. As Cheryl Austin, a community organizer and lifelong resident of Tremé who self-identifies as African American, put it: "Growing up in Tremé, you had a bar, a church, and a funeral home, so you always knew where your family was" (semi-structured interview 2007). For Cheryl, the life experiences and social relations in these sites had shaped Tremé residents into people with unique affective attachments and preferences. Cheryl continued:

> For me, if you did not grow up here, you cannot appreciate living here. . . . For outsiders, the most important things here are the buildings. For us, it is our culture. For us, that is what we consider community, not the buildings. (semi-structured interview 2007)

Cheryl's words resonate with the observation by George Lipsitz (2006) that working-class African Americans, whose spatial mobility is curtailed through everyday forms of apartheid—for example, police profiling, suburban white flight, de facto exclusion from suburban public schools— develop ferocious attachments to those spaces they are relegated to, and these attachments help them make meaningful lives in an otherwise hostile environment.

Lipsitz's commentaries on attachment resonate with Didier Fassin's (2013) claim that people come to experience emotions in particular ways

FIG. 14. Candle Light Lounge in Tremé. Courtesy of author.

depending on the position they inhabit in a society's body politic (the structuring of social and spatial relations in terms of ethnicized, gendered, or racialized difference). Fassin argues that while resentment is often viewed as a morally questionable emotion in contexts such as post-apartheid South Africa, it is important to recognize its value as a means of remembering transgressions people hope not to repeat. Emotions, then, manifest in varying ways and are contingent on the positionality of a person in the web of socio-political relations. As I have continued to reflect on Cheryl Austin's words, I have found Lipsitz's and Fassin's observations to be a helpful point of departure. They both suggest that emotions have historicity, and to understand their historical contingency, we must interpret them in terms of the ever-evolving spatialized and racialized relationships that make up the birth, growth, and life of a place like New Orleans.

In defining what Tremé is, the importance of people's familiar relationships was a key theme that arose during my casual conversations with neighborhood residents. On one occasion, Mr. Lionel Jones told an anecdote about an out-of-state visitor who had seen the HBO television show *Treme*, in which writers loosely used the neighborhood and its cultural practices as inspirations for devising story lines about post-Katrina New Orleans. The visitor quizzed him on the existence of landmarks and personages in the series. When Mr. Jackson doubted the actual existence of such places and people, the visitor challenged his standing as a local expert. Mr. Jackson replied to the visitor's doubts:

> Do you see that man with the hat, the man asking for money for the second line? Do you know his name? Do you know his mother's name? Do you know who his cousins are? I do, and *that's* what it means to know Tremé. (unstructured interview 2010)

Other life experiences that shaped the dispositions and affective attachments of many Tremé residents were the second-line and Mardi Gras Indian parades. As Cheryl Austin does, Mr. Collins Lewis explicitly references the ways that participation in Tremé's unique space-making practices (its ecology of affect) shaped its residents' dispositions and attachments:

> When you live here in Tremé, music is every day. People know this is what goes on. People who weren't born and raised here, they don't understand it. The police don't understand the culture of New Orleans. This is what we do. It's like going to church. When the music sounds, you go dance for four hours. Mainly, we're dealing with something only black people deal with. People from outside of this area, they don't know this is a spiritual thing. (unstructured interview 2010)

During the two decades preceding Katrina, Tremé witnessed the arrival of a new group of residents who were attracted to the neighborhood by its affordable historic properties (Reckdahl 2007). Suburban flight and the proliferation of vacant and blighted properties lowered home values, and given the area's proximity to the city's tourism and business centers, buyers saw lucrative returns on real estate investments. Gaining on these real estate ventures, however, would require transforming the neighborhood to suit the sensibilities of those affluent professionals who had not previously lived in the area. Newly arrived gentrifying homeowners formed a neighborhood association dedicated to the area's cultural transformation (Reckdahl 2007, 2008). Association organizers attempted to curtail second-line parades and opposed the opening and patronage of neighborhood bars. These efforts intensified following Hurricane Katrina, when gentrifying New Orleanians saw the disaster as an opportunity to further displace their working-class neighbors. In 2006 one of the neighborhood association's members wrote on the online blog *New Orleans Renovation*:

> Pre-Katrina, it [Tremé] suffered from a rash of drug violence which has been plaguing the second line scene for the past few years. Joe's Cozy Corner [one of the neighborhood's landmark lounges] was the site of a few murders. North Robertson Street still suffers from some bad elements who persist but we could see that the storm had created a vacuum of leadership and for a while, crime. We decided this was the time to get start [*sic*] a very solid and very professionally managed neighborhood organization that would welcome all residents, erasing the past bad feelings that existed between old, ineffective groups that had failed to make any real positive impact in the area. After Katrina, the blighted ghetto looks more like a slum. The Historic Faubourg Treme organization is dedicated to the smaller historic area of Treme and focused on our main problems of Crime, Blight and Grime. (*New Orleans Renovation* 2006)

Strategies of neighborhood revitalization such as this one ignore the broader and deeper histories of spatialized racism that imposed marginality and poverty on many working-class African Americans in southeastern Louisiana. Rather than proposing to address these processes as a strategy for combating inequity, the preceding blog post accuses established grassroots organizations of inefficiency and makes a simplistic correlation between second-line parades, neighborhood bars, and violence.

Lifelong residents, in contrast, viewed the practices of gentrifying neighbors as potentially ethnocidal. Cheryl Austin, for example, commented:

> We have cultural differences. I like walking out of my house and having a beer. A lot of them think it's bad. . . . That is happening right now, with gentrification. They don't want no bars open, they don't want young men walking around with T-shirts and their jeans pulled

down, they don't want anyone hanging out. These cultural things are beginning to change. People that have been here a short time— and I mean twenty years is a short time, twenty years or less—they want the second lines to clean up after themselves! (semi-structured interview 2007)

Cheryl's words call attention to the cultural stakes of gentrification in Tremé. Relatively recent arrivals who do not have the affinity of lifelong residents to second lines, neighborhood bars, and subaltern uses of urban space (e.g., hanging out) misunderstand the neighborhood and impose their desire for "clean" spaces where neither rubbish nor working-class African Americans are visible.

When Ecologies of Affect Meet Affective Regimes

Prior to the hurricane, New Orleans lacked a master plan for development. Although a plan had been in the works since 1991, it had yet to be completed and approved due to unresolved differences among various stakeholders. In the catastrophe's aftermath, the U.S. government required the city of New Orleans to finalize a development master plan as a condition for the disbursement of federal reconstruction aid. This requirement included the caveat that the city's master plan should also include a plan for recovery, and this supplemental document had to be drafted via a participatory process.

In the succeeding three years, the people of New Orleans observed and partook in the creation of three official recovery-planning initiatives, each succeeding the other. These plans represented different regional political and economic interests.

BRING NEW ORLEANS BACK PLAN

Mayor C. Ray Nagin organized the first of the city's recovery plans, Bring New Orleans Back (BNOB), and commissioned the document's drafting to the Washington DC–based Urban Land Institute (ULI) in the late fall of 2005. The city's most renowned real estate mogul and banker, former ULI president Joe Canizaro (also New Orleans's largest contributor to George W. Bush's presidential campaign), was appointed to the BNOB's land use committee. Bring New Orleans Back proposed that reconstruction efforts bypass the city's most heavily flooded neighborhoods and allow these areas to revert to green space. This recommendation incited an onslaught of resident critiques and protests for how the plan ignored the meanings of (and the race and class-based contestations over) neighborhood space in New Orleans.

When BNOB faced widespread opposition, the New Orleans City Council saw an opportunity to challenge the mayor's central role in the reconstruction. It moved ahead with the development of another recovery-planning process in the spring and summer of 2006—the Neighborhoods Rebuilding Plan.

THE NEIGHBORHOODS REBUILDING (LAMBERT) PLAN

The Neighborhoods Rebuilding Plan was drafted under the direction of urban planners Paul Lambert and Sheila Danzey and would come to be informally known as the Lambert Plan (Krupa 2007b; *Times-Picayune* 2006). This second plan was supposedly designed to fulfill the federal requirement for the release of recovery funds in Louisiana that stipulated the recovery plans had to be devised via a public participation process, which BNOB lacked. Speaking about the planning process to a group of engaged New Orleanians at a Neighborhoods Planning Network meeting in June 2006, City Councilwoman Shelley Midura and Councilman Arnie Fielkow described the Lambert Plan as "a process that is bottom up, and not top down," and that "is very important in order to indicate what you [residents] want the directive to the planners to be" (fieldnotes 2006). The council members' words articulated what became a common trope in official representations of participatory recovery planning: the process was a mechanism of shared governance where residents of all socioeconomic backgrounds could act as potential intellectual authors of the city's recovery plan.

Official representations of the Lambert planning process did not match the experiences of many Tremé residents. Rather than a participatory process where attendees had the opportunity to devise a vision of neighborhood recovery alongside expert planners as equal partners, hired architects worked in seclusion for the first three months of the planning process and spoke only with those stakeholders who had been carefully vetted by city planners, thus effectively excluding subaltern

New Orleanians. Once a document was drafted, Lambert Plan organizers called a number of public meetings across New Orleans to present their ideas to city residents.

In Tremé, Lambert and Danzey subcontracted the recovery-planning process to the Miami-based architecture firm Zyscovich Inc. On July 13 Bernard Zyscovich presented his plan for the neighborhood's recovery in St. Augustine Church. The meeting was held in the church's reception area, an old, large room with high ceilings and thick plaster walls. Residents of varied socioeconomic backgrounds including schoolteachers, car mechanics, city workers, and doctors filled the room. Although the Zyscovich team had spent three months in the city, it had consulted none of the attending New Orleanians while drafting the near-final plan.

When Midura and Fielkow emphasized the Lambert Plan would be created through a participatory process where all city residents could partake in its authorship, they failed to mention a couple of caveats. One was the Housing Authority of New Orleans (HANO) and the U.S. government's Department of Housing and Urban Development (HUD) had made a priori decisions for certain urban spaces—namely, Planning District 4's public housing units. HANO and HUD had been in the business of dismantling public housing facilities for several decades, especially after desegregation, when public housing units came to be seen as spaces of "black poverty" in the imagination of many New Orleanians (Breunlin and Regis 2006). The storm created favorable conditions for the opportunistic acceleration of institutional and social time on HANO and HUD's part, and Zyscovich's planning team accepted the agencies' decision to shut down and redevelop major public housing facilities in the area without the public's input. Among these facilities was Lafitte, a project that adjoined the historic section of Tremé and that subaltern New Orleanians considered a part of their neighborhood.

Rather than focusing on facilitating the expedited return of displaced public housing residents to the city, the Zyscovich plan instead proposed a number of projects meant to enhance the area as space of capital investment. One idea called for removing the interstate overpass that had been built in the 1960s, divided the neighborhood in two, and negatively impacted its business district. The project's lead architect introduced the plan, saying:

We're going to bring up a pretty radical idea. We want to reestablish the historic Claiborne Road with a boulevard, sidewalks, and tree planting. We are looking at the idea of bringing down parts of I-10. The idea is to bring people down into a beautiful boulevard. (fieldnotes 2006)

Behind the architect, a portable projector displayed a watercolor-like image of North Claiborne Avenue's possible future (fig. 3). The image showed a wide boulevard, trees planted in sidewalks, and boutique retail spaces. Buildings and trees were rendered in vivid colors while people were represented as generic and interchangeable white silhouettes. The rendering of people as white outlines implicitly conveyed the idea that, for this planning team, the specific identities and embodiments of the neighborhood's residents were not a focal concern of recovery plans. It did not matter whether Tremé's pre-Katrina residents were part of the neighborhood's future population as long as the removal of the interstate ramp created the aesthetics and the workings of a capitalist consumer society. Who could resist the allure of such an affective regime?

On that humid summer night, it seemed the overwhelming majority of Tremé residents in attendance could. One neighbor, Mr. Felix Smith, spoke up. His tone conveyed that the plan had struck an emotional nerve for him:

Armstrong Park was a plan too! One hundred and sixty-four families were moved out. The plan never worked. Your proposed plan, the green space, everything sounds like the entire area is going to be commercialized! I know there are a lot of homeowners in this room. Nobody had any input as to what our neighborhood is going to look like. (fieldnotes 2006)

Mr. Smith's comment opened up a deluge of critical comments on the part of attending residents. They expressed frustration and aggravation at the prolonged closure of Lafitte. One resident spoke up:

Bring the people back that want to come back home. Open up the projects, let the people back in! Everything can't become a green space. People need housing, people need where to live! (fieldnotes 2006)

While the residents made it clear that what they cared about in post-disaster recovery was the reconstruction of their neighborhood's social fabric, expert planners did not record their voices in accordance with official representations of recovery planning. Instead, Zyscovich planning team members attempted to legitimize their vision of recovery by invoking logics of capital investment as uncontestable facts of disaster reconstruction. When confronted with similar critiques of the same plan at a later date, Zyscovich explained:

One of the things we hope to get out of this is the creation of guidelines, sustainable from an energy perspective, so that it's easy to maintain the neighborhood, the overall neighborhood is encouraged. *Private investment is important. When we start getting the money into the city, the overall investment will happen at a faster pace.* Commercial corridors can create a much better neighborhood. Cafés, small grocery stores, drugstores, things that service the community—allow for these things to pop up. As Lafitte gets reconstructed, if we think about planning holistically as an integrated process, *you can get more value out of the dollar.* (fieldnotes 2006, emphasis added)

When residents pressed him further and insisted on the importance of opening up Lafitte as soon as possible rather than prolonging the displacement of the city's most economically vulnerable residents, Zyscovich replied in distinctly neoliberal terms:

Recovery plans need to be sold in terms of their investment potential. The federal government is much more willing to invest five dollars when it is going to get twenty-five dollars in return than [invest] five dollars in mere social services. (fieldnotes 2006)

The exchanges between expert planners and New Orleanians that I observed in post-Katrina recovery planning gave me a glimpse of what subaltern Tremé residents cared about in disaster reconstruction. These residents felt strongly about the relationships among people who had lived in the city before the catastrophe, were familiar with one another, and had forms of personhood that were shaped by their life histories in the city's unique, socially produced spaces: quotidian practices of hanging out, frequenting neighborhood bars, and participating in second-line and Mardi Gras Indian parades. Consequently, these New Orleanians defined recovery as the reinstatement of the area's pre-Katrina population.

Expert planners, on the contrary, prioritized the conceptualization of the neighborhood as a space of capital investment and reproduction, as evidenced in their points: "Recovery plans need to be sold in terms of their investment potential," "commercial corridors can create a much better neighborhood," and "you can get more value out of the dollar." In this affective regime, it did not matter whether New Orleans was inhabited by the same people who had lived there before the storm as long as the people who now populated the city would circulate through its economic corridors and their practices of consumption would realize the area's neoliberal potential (i.e., stranger sociality). These latter visions of disaster recovery featured the prescription of spaces associated with urban renewal—"cafés, small grocery stores, drugstores"—that were assumed to produce affective reactions of pleasure and desire among planning process participants.

The planners' narratives also used accompanying imagery that emphasized both the use of color and the texture of architectural forms to create a sense of novelty and cosmopolitan style (fig. 3). The mobilization of this imagery of recovery planning is what I attempt to grasp through the concept of affective regime, especially as it relates to the way social actors in positions of political or institutional power make decisions about how space and reconstruction resources are used based on hegemonic notions of what is pleasant, desirable, and conducive to well-being.

THE UNOP PLAN

In the late summer of 2006, a third participatory planning process titled Unified New Orleans Planning (UNOP) sidelined the Lambert Plan. The BNOB and Lambert Plan represented attempts by different sectors of city government—the mayor's office and the city council, respectively—to exercise some control over the reconstruction of New Orleans. Now UNOP constituted an attempt by Governor Kathleen Blanco's administration and the Rockefeller Foundation to gain influence in the city's recovery. Its organizers promoted it as the road to the city's definitive plan.

The UNOP process, which was organized on the ground through a partnership between the local urban planning firm Concordia and the development non-profit the Greater New Orleans Foundation, involved several town hall meetings and community "participation" activities in the fall of 2006. On December 16, 2006, District 4 residents gathered at a high school on Esplanade Avenue (one of the city's historic thoroughfares) to view the presentation of the nearly final plan. Architects from the St. Louis–based architecture firm HOK presented the document to an audience of nearly sixty attendees. The proposal focused on one image, which pictured New Orleans from an aerial perspective and contained superimposed intersecting translucent red arrows (fig. 15). The arrows linked landmarks such as Louis Armstrong and City Parks to the city's tourism center, the French Quarter. One of HOK's architects presented the image, saying:

> What is the potential for Louis Armstrong Park? Historically, it is very significant. Louis Armstrong is one of the most important elements of New Orleans. Armstrong must be connected to the river and to City Park. We need to consider its relationship to Jackson Square and to Iberville. . . . Tremé needs to be integrated into two corridors, one with the French Quarter and Lafitte, and one with Iberville. (fieldnotes 2006)

In addition to its intersecting arrows, the plan also included four polygonal spaces differentiated from the rest of the urban matrix through translucent yellow highlighting. These polygonal spaces were drawn directly over four of the city's major public housing structures, which HUD and HANO had ordered closed immediately after the storm: Lafitte, St. Bernard, C. J. Peete, and Iberville. Although these public housing facilities had only sustained light damage and were by and large inhabitable, they were scheduled for demolition, reconstruction, and privatization to accommodate private businesses such as film studios and a variety of mixed-income housing developments. In essence, the plan proposed that the recovery of the city's central neighborhoods would best be achieved through creating spatial relationships between architectural structures that facilitate the movement of people and capital across New Orleans, hence the red arrows conveying the idea of the city as a mechanism of capitalist circulation. In this vision of recovery, public

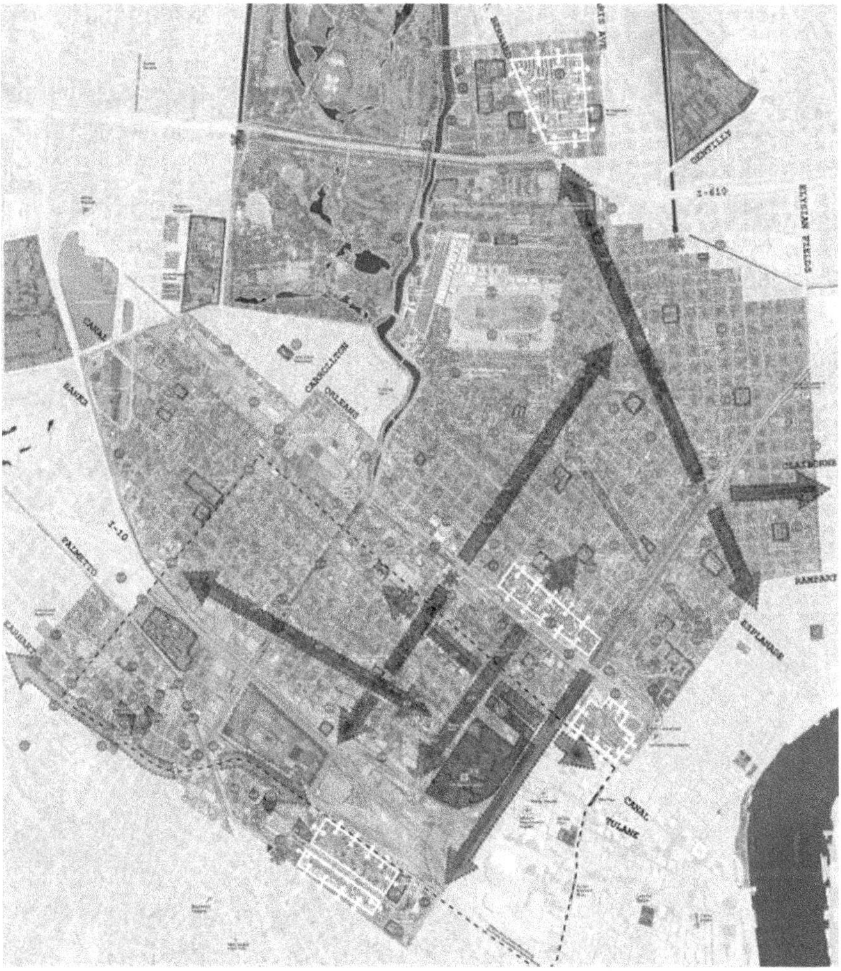

FIG. 15. UNOP recovery plan for Planning District 4. Created by Frederic Schwartz Architects. Courtesy of New Orleans City Planning Commission.

housing structures were deemed an impediment to this circulation (and replication) of capital; therefore, they were scheduled for removal and the areas slated for "revitalization."

As with "Flood Plain Management" and Limón de la Cerca's master plan, the UNOP image omitted people as a function of its scale, but this omission did not preclude the plan's focal figure from articulating assumptions about the nature of human beings, community, and the relationships between people and things that its creators considered necessary to produce recovery. The people who would one day inhabit this imaginary landscape would relate to one another as strangers and be brought together by the mechanistic workings of the city as a space of capitalist circulation.

New Orleanians attending the meeting quickly picked up on the plan's assumptions. The overhead image of District 4 evoked heartfelt critiques. Residents claimed that, as presented, UNOP ignored what they considered to be the most important issue: the return of Planning District 4's pre-Katrina population who were not strangers but neighbors with names, histories, shared experiences, and specific kinds of sensibilities and practices.

One resident, Mr. Eubanks, spoke up. His voice carried a sense of aggravation with the allegedly "participatory" planning process that seemed to have silenced a number of voices:

> Two things. One is about affordable housing. Again and again, it's been stated here in District 4 that we are very much afraid of what happened in St. Thomas. Two thousand families were removed and only about a hundred families returned. There have to be some things right now that say: how do we get people, who want to come back, home now? We have discussed a number of ways of getting existing homes ready for people to come back. Somehow, something has to be put in these plans. They are renters. We need to get these people back. Where is the housing issue? The most important issue? (fieldnotes 2006)

Mr. Eubanks made a reference to St. Thomas Development, one of the city's major public housing projects, which had been demolished and redeveloped as mixed-income housing prior to Katrina. In the same way, the UNOP planners proposed to redevelop the four public housing projects in District 4. Mr. Eubanks invoked St. Thomas because the redevelopment project, begun as part of HUD's Hope VI initiative, was so egregiously mismanaged by the Housing Authority of New Orleans that it experienced extensive delays, displaced the housing area's residents, and eventually required federal interference for its completion. The project's mismanagement forced many of the collective housing unit's residents to seek alternative living arrangements, and they never returned to the area. Mr. Eubanks was wise enough to know that urban development plans in New Orleans seldom unfold as advertised and that when it comes to residents of public housing, their dispersion and loss as members of an urban community are not regretted. His comment addressed precisely those yellow polygons that HOK architects had not mentioned but that weighed heavily on the minds of many Planning District 4 residents who objected to the closing and redevelopment of public housing. Mr. Eubanks and the other residents insisted that the redevelopment would prolong the displacement of working-poor New Orleanians and could very well prevent their return. Mr. Eubanks may not have been entirely incorrect; by 2013, 111,000 African American residents of New Orleans had not come back to New Orleans (Mack and Ortiz 2013).

Mr. Eubanks's tone conveyed a sense of frustration, but it did not compare to what followed. From the back of the room, a former resident of public housing angrily spoke up: "There ain't gonna be no motherfucking planning until you open up the projects and bring everybody back!" (fieldnotes 2006). This statement initiated a wave of complaints about the plan, while the presenting architects stood on the school's stage with lips pursed in what appeared to be condescending smirks. After a few minutes of heated comments, police were called into the room, and officers removed the protesting residents from the building. The remaining attendants had mixed feelings about the summoning of police. Even those who were struggling to understand the vision behind the UNOP plan and were willing to give its creators the benefit of the doubt felt the forceful removal of the protestors was unwarranted. After a few tense moments, though, the meeting proceeded, and the planning team completed its presentation.

Later, planning team members hung poster-size versions of the plan on the school's walls for the residents to examine. As I stood by a ten-foot version of figure 15, I saw participants walk up to the image and gaze at it, intrigued, as if attempting to make sense of what was presented to them as the state of the art in urban development. Participatory planning, although heralded by city council members as the collective authorship of the city's reconstruction by New Orleanians, seemed more akin to a pedagogical process where expert planners attempted to teach participating residents new ways of seeing and affectively experiencing urban spaces.

Attachment and the Subaltern as Costs of Neoliberal Recovery

In this chapter, I have shown how the dispositions and attachments of Tremé residents—what they cared about and how they cared about it in disaster recovery—took shape over life experiences in the socially produced spaces and socialities of what I call New Orleans's ecology of affect. Moreover, I demonstrated that these spaces and social relations are by no means generic or value free; they took form through discourses and practices of racial difference making over the course of the city's three-hundred-year history. I have also shown how participatory recovery-planning processes, rather than documenting the voices of city residents and converting these voices into a reconstruction directive, became mechanisms by which expert planners attempted to instruct city residents on novel ways of making, experiencing, and appreciating urban space. In this affective regime of recovery planning, hired experts ignored the attachments and sensibilities of subaltern New Orleanians and promoted the spatial arrangements (commercial corridors conducive to capital investment, circulation, and replication), aesthetics (renovated spaces formerly occupied by public housing), and socialities (a city populated by strangers) of early twenty-first-century capitalism as uncontestable facts of disaster reconstruction. Recovery-planning meetings, rather than being forums for recording residents' visions of the city's future, turned out to be tempestuous spaces where ecologies of affect met with affective regimes.

While I was conducting a public presentation of this chapter at a workshop on disaster mitigation in Chiapas, Mexico, in 2013, a participant asked, "So what if that's how some people in New Orleans feel about their neighborhoods? What if neoliberal approaches to disaster reconstruction actually work?" I found this inquiry interesting because it dismissed affect and emotions as superfluous aspects of disaster recovery in favor of economic and infrastructural indicators. The question's disregard for the affective dimensions of disaster reconstruction reminded me of Andrew Pickering's (2008) comment in the aftermath of Katrina's devastation that New Orleanians had to let go of attachment and simply resettle the city to a location not fraught by flood risk.

To simplistically recommend that New Orleanians move to areas not threatened by flood risk without recognizing the relationship between spatial production, racialized differentiation, and affective experience is not only insensitive but also unrealistic as a course of action. I am not suggesting that we should ignore the potential plight of people who live in disaster risk zones, but in places such as New Orleans, we must first understand and address the space and difference-making practices that simultaneously limit people's socio-spatial mobility, distribute flood risk inequitably, and shape the ways people come to emotionally relate to urban space. Resettlement alone will not solve the tensions and inequities engendered in New Orleans's body politic, and these tensions and inequities must first be addressed if expert planners ever hope to secure the assent of New Orleanians to relocate their city and neighborhoods.

Over the course of my experiences in recovery-planning meetings after Hurricane Katrina, not once did expert planners mention the issues of space and race. To suggest that New Orleanians "let go of attachment" without addressing these key dimensions of the city's urban landscape is to presuppose that space is a neutral backdrop of social action where all people, regardless of the way they are racialized, can move freely and equitably. Nothing could be more out of touch with the lived experiences of many New Orleanians.

Recently, Gastón Gordillo (2014) observed that the production of space always involves the destruction of something else, be it either a forest that is leveled in the name of new construction or indigenous populations that are massacred or displaced in the formation of colonial and national societies. Consequently, Gordillo calls for a critical interrogation of capitalist space, which official state narratives often celebrate as a positive and unquestionable development. These observations are fitting in the case of New Orleans, where neoliberal recovery plans proposed the realization of a capitalist space as a means of recovery but at the cost of many subaltern lives, both social and biological (Adams 2013). Expert planners who work on recovery-planning initiatives should reflect on the ethics of their practice and their obligation to those members of a population who are most vulnerable to the social and environmental effects of disasters. In New Orleans, given that those most impacted by Katrina's effects—low-level service sector workers, public housing residents, the working poor, and socioeconomically marginalized African Americans—were not deemed focal "clients" of recovery plans, it is no wonder that their displacement from the city was prolonged and that state agencies abdicated responsibility to look after their well-being.

6. How to Care?

Before Hurricane Katrina few people outside of New Orleans had heard
of or cared about the Lower Ninth Ward, and few New Orleanians who
did not live there visited the neighborhood. Residents of the Lower Nine,
as the neighborhood is colloquially known by New Orleanians, were
painfully aware of the ways their home was stigmatized by associations
with crime and poverty—a stigma that also carried racist undertones.
Over the course of my ethnographic work in this area since 2008, residents have repeatedly mentioned the impact of the neighborhood's
unfavorable reputation on their lives. I distinctly remember Jeanell
Holmes, a Lower Ninth Ward resident I met during this project, once
saying, "My friends in high school wouldn't come to pick me up or drop
me off when we went out" (unstructured interview 2009). The words
of Victoria Jackson, another resident and community organizer, also
stand out in my memory: "The city always looked at us as a downtrodden neighborhood" (fieldnotes 2008).

New Orleanians did not always see the Lower Ninth Ward negatively.
Located less than four miles east of the French Quarter along the Mississippi River (fig. 16), the area was first settled in the 1860s by immigrants from various parts of Europe, including Ireland, Germany, Italy,
and by emancipated African Americans who established small family
farms (GNOCDC 2007). In the early twentieth century, during the era of
state-sanctioned segregation, the neighborhood was purposely developed as a place where working-class African Americans of modest means
could build and own their own houses.

Like other neighborhoods of New Orleans, the Lower Ninth Ward
remained a socioeconomically diverse area until the 1960s, when the
era of suburban flight and de facto segregation initiated a process of

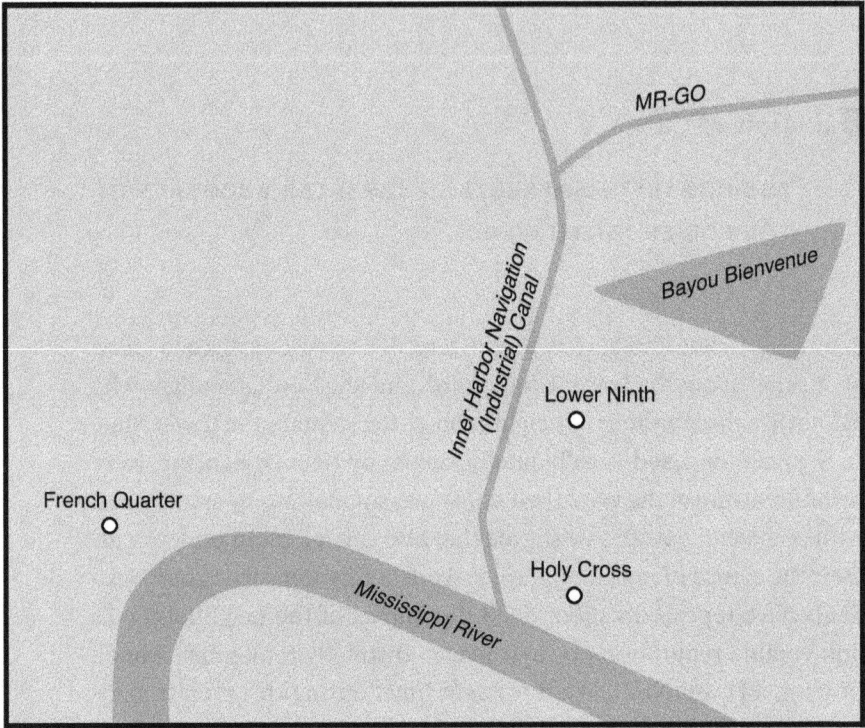

FIG. 16. Lower Ninth Ward area. Courtesy of author.

racial and economic polarization in the city. By the time Katrina reached the U.S. Gulf Coast, 95.3 percent of the neighborhood's nineteen thousand residents self-identified as African American in the U.S. census, and the median household income was $20,000 per year (GNOCDC 2007). Despite the financial difficulties experienced by many families, Lower Ninth residents prided themselves on a high homeownership rate of 59 percent (compared to 46.5 percent in New Orleans Parish) and on a history of economic self-sufficiency. Recollecting a conversation she once had with a fellow New Orleanian from a more prestigious neighborhood, Jeanell once told me the words she used to express this ownership pride to non–Ninth Ward residents: "Y'all rented, but we [Lower Niners] owned" (unstructured interview 2009).

Despite its proximity to the French Quarter, the experience of many people from the Lower Ninth Ward was one of relative isolation from the city's financial and symbolic centers. In the early twentieth century, the U.S. Army Corps of Engineers constructed a navigation channel and wharf known as the Industrial Canal (officially titled the Inner Harbor Navigation Canal), which cut the neighborhood off from easy access to the rest of the city (fig. 16). In the mid-century, this canal was directly connected to the Gulf of Mexico via another human-made navigation channel, the Mississippi River Gulf Outlet. The resulting saltwater intrusion had a significant environmental impact on surrounding wetlands and devastated nearby cypress forests and bayous. During and directly following Hurricane Katrina, multiple levee failures along the Industrial Canal led to rapid flooding, which caught those families who would not or could not afford to evacuate by surprise. At least seventy-five people died. The news media used images of the destruction caused by the levee breaches to represent Hurricane Katrina to the rest of the world, and the Lower Ninth Ward went from being largely unknown to becoming a household name overnight (Breunlin and Regis 2006).

Prior to the disaster, not only was the Lower Ninth Ward stigmatized by many New Orleanians, but it also existed at the margins of the city's political culture. Many of the city's power brokers either grew up, lived in, or identified with areas such as Uptown, the Seventh Ward, or New Orleans East, and they manifested identities or alliances that were nuanced by the city's racial identity politics. While the city council and state congressional candidates historically viewed the Lower Ninth Ward residents as an important voting constituency, the city government had a history of chronically underserving the neighborhood. The government poorly maintained or developed all of the ward's roads, sewage, and levee systems prior to the storm, as it considered the area secondary in comparison to better-off parts of the city and its commercial and tourist districts.

The devastation caused by Hurricane Katrina and the neighborhood's rise to prominence via news media coverage, however, placed the Lower Ninth in what its residents still see as a somewhat paradoxical situation. The neighborhood's destruction has become iconic of the disaster, and many people in the Lower Ninth believe that the imagery of their suffering provided much of the emotional leverage that raised national sympathy for New Orleans. Despite this newly attained iconic status, civically involved residents feel the neighborhood continues to exist in a marginalized relationship to city government and reconstruction resources years after the flood. For example, Lower Ninth Ward residents systematically received lower compensation from the Louisiana Recovery Authority's Road Home Program for the damages their houses suffered, because it used pre-Katrina property values to estimate the homeowners' losses (Burdeau 2011). These pre-Katrina figures, in turn, were the product of spatially deployed racist and classist prejudices that shaped property values in pre-Katrina New Orleans, and by using them, the program perpetuated racially based inequalities.

Post-Katrina the Lower Ninth Ward has also continued to exist in a marginal space in relation to city government. In 2008 the Office of Recovery and Development Administration, which Mayor C. Ray Nagin created and Edward Blakeley directed, scheduled twenty-two recovery projects for the minimally damaged area of Uptown, where the city's premier elite neighborhoods are located. One of the projects included the construction of tennis courts. Meanwhile, the Lower Ninth Ward was scheduled to receive only three recovery projects (ORDA 2008).

These policies took place within a broader context of neoliberal disaster reconstruction (Adams 2013; Barrios 2011; Gunewardena and Schuller 2008). For example, the flagship project of the Office of Recovery and Development Administration (restructured as the Office of Community Development in 2009; see Eggler 2009) featured the creation of seventeen "target recovery zones" that were supposed to be the foci of the recovery effort in New Orleans (Marszalek 2007). The plan proposed using eminent domain laws and reconstruction funds to create nodes of investment (i.e., shopping malls, film studios) in these zones that one day would produce the tax revenue to fund the public services many New Orleanians urgently need. In the meantime, critical public services such as the city's Charity Hospital, which was a regional resource for economically marginalized Louisianians, remained closed and without a replacement seven years after the storm (*Times-Picayune* 2011).

I regard the seventeen-target-zone recovery plan as neoliberal because instead of immediately using public resources to provide the services (firehouses, schools, police stations, public health clinics) that struggling middle-class, working-class, and working-poor New Orleanians needed to make the city habitable, the plan proposed using these resources to encourage private, out-of-state investment under the legit-imization that such investment would one day produce the tax revenue necessary to fulfill the aforementioned public needs. The free market buttressed by the state, this plan suggested, would provide the means to social well-being.

I recognize that a critical reader may object to my use of the term *neoliberalism* and argue that what we are seeing here is a state-supported, private sector type of capitalism. A pure neoliberalism, in contrast, would involve completely removing the state from reconstruction efforts and leave the free market to sort them out. My argument in support of using the term is that a pure neoliberalism, like a pure capitalism, has never existed. Instead, what we are seeing here is the manifestation of one of a variety of neoliberalisms that take form as narratives concerning the application of capitalist cost-benefit analysis to all facets of human life take hold of the imagination of government officials, urban planners, and policymakers in post-disaster reconstruction.

The neoliberal approach to reconstruction has created a gap in the provision of public services and reconstruction efforts in the Lower Ninth Ward. Consequently, a number of academics, environmentalists, and non-profits descended on the neighborhood with the intention of addressing the shortcomings of capitalist disaster reconstruction. This phenomenon is epitomized by actor-turned-philanthropist Brad Pitt who, upon arriving in the Lower Ninth Ward and witnessing the neigh-borhood's slow pace of recovery, coined the phrase "Make It Right." This phrase became the name of a non-profit organization designed to help families obtain low-interest mortgages to finance the construction of energy-efficient and environmentally friendly homes.

All newly arrived organizations and experts are interested in making a difference in the neighborhood's reconstruction, but all are driven by their own agendas and definitions of well-being and recovery. Furthermore, the various agencies, institutions, academics, and concerned citizens vary in the degrees to which they have been able to negotiate their visions of recovery with neighborhood residents, who themselves form a heteroge-neous group that speaks with multiple voices. Keeping these complica-tions in mind, this chapter explores the implications of the different ways reconstruction actors—environmentalists, academics, residents of varied socioeconomic backgrounds, non-profit workers, philanthropists—care about the Lower Ninth Ward in the aftermath of Katrina. Specifically, I ask: What do reconstruction actors care about and how? What is at stake in the different ways they care about the Lower Ninth?

Over the course of ethnographic research I have conducted since 2008, I have learned that answering my questions requires careful attention to the historical political ecology of this part of the city, and this observation once again brings home the idea that affect has ecology and historicity. As I outline in succeeding sections, even before Katrina, the Lower Ninth Ward was a socially complex space where residents developed varying identities and space-making practices. These differences did not vanish in Katrina's aftermath; rather, they came to matter in new ways in the context of a popular movement to "rebuild sustainably." To further complicate matters, city officials have exploited the multi-vocality and identity politics of the area as a pretext to withhold reconstruction resources from the neighborhood while they claim that, without consensus, recovery cannot move ahead. I hope that by reviewing the historical political ecology of affect, space, and identity in the Lower Ninth Ward, this chapter can help reconstruction actors negotiate their affective sensibilities and priorities with disaster-affected populations in a way the latter find just, meaningful, and relevant.

I collected the evidence presented here over a period of six years, from June 2008 to July 2014. This period is critical because disasters are phenomena that often attract much media and even scholarly attention during their "emergency" phase, but interest in specific disaster-affected localities tends to wane until landmark anniversaries occur one or two decades later. Ethnographers, in contrast, are inclined to maintain long-term contact, providing powerful analyses of the everyday challenges of a disaster's recovery and the lives of the people who live through it. My research consisted of making observations and documentation of neighborhood association meetings and neighborhood social events and of collaboration with grassroots environmental restoration programs. I also conducted ethnographic interviews with neighborhood organizers who represent various Ninth Ward constituencies. Over the course of this work, I developed long-term relationships with a select group of interlocutors who have shared their friendships, private lives, and difficulties with me.

The Historicity and Ecology of Caring

The Lower Ninth Ward gets its name from voting districts used for city elections first delimited in 1809. The Ninth Ward itself was not designated until 1852, when city officials demarcated its limits along Almonaster Avenue; Lake Pontchartrain, the St. Bernard Parish boundary; and the Mississippi River. Between 1918 and 1923, the construction of the Industrial Canal divided the area in two, creating a separation of what became known as the upper (up river) and lower (down river) sections of the Ninth Ward (Campanella 2006). The Industrial Canal was built with the intention of providing commercial ships a shortcut to the Gulf of Mexico, allowing travel between the Mississippi River and Lake Pontchartrain, and a deepwater wharf space. While the canal proved a great financial benefit to the Port of New Orleans, cutting costs by expediting travel and increasing profits through its additional wharf space, its construction also exacerbated the neighborhood's flood risk by bringing massive amounts of lake and river water into its vicinity (fig. 16).

The political ecology of the Lower Ninth Ward was further complicated in 1958, when the U.S. Army Corps of Engineers began the construction of yet another major canal, the Mississippi River Gulf Outlet, which expedited the movement of cargo ships between the Port of New Orleans and the Gulf of Mexico (Campanella 2006). The canal had a number of unexpected effects on the neighborhood's surrounding wetlands: it changed their salinity levels, which in turn destroyed the surrounding cypress forests that had served as a buffer from tropical storms, thus granting storm surges a pathway into the area.

Like many other residential parts of New Orleans, the Lower Ninth Ward was more socially diverse during its early years than at the turn of the twenty-first century when Hurricane Katrina struck. In 1960 the city of New Orleans as a whole had a population of 627,525, with 233,514 (37 percent) of these residents self-identifying as black in the U.S. census. By 2004 the city's population had declined to 462,269, and the percentage of residents who self-identified as black increased to 68 percent (Campanella 2006; GNOCDC 2007). These citywide figures represent a complex process of urban flight that had multiple motivations, with racism being a principal one. As discussed in chapter 5, after state-sanctioned segregation ended in the 1960s, those New Orleanians who self-identified as white moved to the city's western suburbs in Jefferson Parish and the eastern city of Chalmette in St. Bernard Parish. At the same time, many middle-class African Americans moved to the eastern suburbs of New Orleans East and Pontchartrain Park. Many African Americans also left the greater New Orleans area in search of employment opportunities in states where racial discrimination was not as prevalent in the professional job market (Jackson 2011).

It is noteworthy that these patterns were general tendencies and not necessarily an absolute state of affairs. Part of the challenge of discussing the role of space in producing and maintaining racialized identities in New Orleans is that not all city residents engaged in suburban flight. New Orleans today is also characterized by proximate residences of and social relations among people who have diverse genealogies and complex ways of self-identifying. Areas of the city such as Uptown feature neighborhoods whose inhabitants vary in income level, occupation, and identity claims. For example, in Uptown, low-income families live in the backstreet parts, while the areas surrounding Audubon Park and lining St. Charles Avenue feature the most prestigious housing in the city. Nevertheless, the general shifts in the city's demography following 1960 did take place, and they had significant effects on the city at large, such as the proliferation of blighted properties due to abandonment, a decrease in tax revenues to support the city's infrastructure and public services, and the diminished social and spatial mobility of African American residents of modest means.

The racial motivations behind New Orleans's urban flight were also nuanced by the meanings of life in suburbia (the imagined escape from inner-city "problems," the association of modernity with suburban life) and by federal subsidy programs meant to encourage home ownership that drew city residents to outlying suburban areas (Schuller and Thomas Houston 2006). Broader economic issues such as the city's financial crisis during the 1980s and the downscaling of oil company operations also had an impact on New Orleans's population, leading to its continued decline (Sorant, Whelan, and Young 1984).

By the turn of the twenty-first century, the Lower Ninth Ward had witnessed a process of racial polarization. The urban flight that created this polarization was accompanied by de facto segregationist practices meant to limit the spatial and social mobility of working-class and socio-economically marginalized African Americans, and the Lower Ninth became a neighborhood stigmatized as a place of poverty and perceived criminality. But the Lower Ninth Ward was more complex on the eve of Hurricane Katrina than these representations would lead one to believe. As noted, the general area of the Lower Ninth was divided into two distinct neighborhoods, lending yet another dimension to the complex identities of the New Orleanians who lived there. The area's southern edge, which extends from St. Claude Avenue to the Mississippi River, is recognized as the Holy Cross neighborhood by residents and city officials, while the area extending northward from St. Claude Avenue to Bayou Bienvenue (a prominent wetland area) is simply referred to as the Lower Nine.

In the year 2000 the Lower Nine counted a population almost three times as numerous as that of Holy Cross (14,008 versus 5,507 people, respectively) with the former having a significantly higher percentage of residents who self-identified as African American (98.3 percent versus 87.5 percent). Holy Cross also had a much higher percentage of people who self-identified as white (9.4 percent versus 0.5 percent in the Lower Nine). In economic terms, Holy Cross had a slightly smaller number (48 percent versus 50.4 percent in the Lower Nine) of households reporting a low annual income (less than $20,000 per year) and double the amount (3.3 percent versus 1.6 percent in the Lower Nine) of households reporting earnings exceeding $100,000 per year (GNOCDC 2007). While socioeconomic differences between the two neighborhoods might seem minimal from an outsider's perspective, neighborhood identities and varying space and personhood-making practices came to matter in significant ways before and after the storm for people on both sides of St. Claude Avenue.

In 1990, city officials declared Holy Cross a historic neighborhood, and this part of the Lower Ninth Ward saw an increase in new arrivals who were attracted to the area by low property values and historic homes. In this time, Holy Cross residents formed a neighborhood association that rose to prominence as a grassroots-organizing powerhouse after Hurricane Katrina. Although the two neighborhoods had broad socioeconomic similarities, many residents felt some Holy Crossers thought of themselves as distinct from (and perhaps even slightly superior to) their counterparts in the Lower Nine.

Holy Cross neighborhood organizers and residents accused of subtle forms of elitism feel misunderstood and misrepresented when confronted with the sentiments of their fellow Lower Niners. But some Holy Crossers did (and continue to) engage in quotidian practices of spatial production and spatial use that differ from those of other residents and that have long histories of creating racialized class differences in New Orleans (Barrios 2010; Regis 1999). These practices include avoiding or expressing disdain for neighborhood working-class bars, gentrifying historic neighborhoods, and monitoring working-class residents' use of public street spaces for daily socialization. As Setha Low (2000, 2003, 2009, 2011) has commented, neighborhood associations have the potential to articulate subtle forms of spatial segregation through the seemingly innocuous act of "making things nice." Pierre Bourdieu (1977), in turn, would recognize the professionalization of neighborhood life through institutions such as neighborhood associations as a key element of bourgeois habitus—that is, the ways people give form to space and time through the deployment of structured social relations and the ways people come to experience their sensibilities and embodied dispositions over the course of life experiences in these spaces and temporalities.

Post-Katrina, Lower Ninth Ward residents have continued to struggle with the subtle differences in habitus and neighborhood identity that differentiated the Lower Nine from Holy Cross, differences that have a bearing on how residents care about the area's recovery. These struggles stem from the opinion of some Lower Niners that Holy Crossers have a long history of looking after their neighborhood's self-interest rather than that of the Lower Ninth Ward as a whole, and they are active in securing resources from city government and philanthropic organizations and then not sharing them with other neighborhood organizations. These tensions among Lower Ninth Ward residents present a significant challenge to all neighborhood organizers, regardless of which constituencies or neighborhood identities they sympathize with, because they agree that a broad-based constituency is more efficacious at making demands before local government officials for recon-struction resources.

Resident organizers of various walks of life are also overly aware of the ways city government officials (city council members in particular) exploit the neighborhood's complexity and multi-vocality, insisting that reconstruction projects are not allocated to the neighborhood because of residents' "incapacity" to speak with one voice. This excuse masks the historic stigmatization and marginalization of the Lower Ninth Ward in city politics. In the post-Katrina moment, neighborhood organizers have conflicting opinions on how differences in identities, habitus, and visions of recovery should be negotiated.

Discussions of these varying positions routinely involve the use of proxies for racialized class difference. Take, for example, a conversation among neighborhood organizers from Holy Cross and the Lower Nine that took place in June 2008. The organizers had gathered at a French Quarter restaurant to hold a daylong conversation about how to better coordinate among neighborhood groups so they could present a unified political front before the New Orleans City Council, City Planning Commission, and the mayor's office. As they discussed these topics, John Jackson, a resident organizer from Holy Cross, shared his views on neighborhood identity politics with the dozen or so other neighborhood organizers at the table:

> Once we rid ourselves of traditional thinking, we can move on to cre-ating the future. I say a lot of times, if you hold onto the past, you can't move on to a brighter future. Not to say you don't look back to the past to get some wisdom. But I know there has been bickering in the Lower Ninth. But Katrina has given us an opportunity to build a better city. I mean, slavery was once a tradition, but we let go of it. (fieldnotes 2008)

In this instance, Mr. Jackson also used a temporal metaphor to speak of habitus and identity dimensions of neighborhood life, and he divided residents of the Lower Ninth Ward between those who looked "forward" and those who harped on "past differences." This temporal language attempted to dismiss what remained serious concerns in the neighborhood, such as gentrification and elitism, and discredited those who took up such causes as resentful obstructionists.

His comment elicited a response from Bernie Holmes, a respected grassroots neighborhood organizer from the Lower Nine, who said:

> I am a traditionalist, because I look to the past to see if they have any traditions to make the future better. I think we need to look at the past to see what is going on. Because games are still being played. The money that is coming to our city, how is it being misused? This city is getting millions and millions of dollars, but we don't get it. The majority of the money this city is getting is on the back of the Lower Ninth Ward. That's where the collaboration has to happen. . . . [Before Katrina] the Lower Ninth Ward provided its own economy. When my dad lived there, a lot of people who lived in the Ninth Ward survived on the Ninth Ward; they didn't have to leave. Now I don't know what's being planned, but they're trying to take it away. You talk about segregation, we have segregation within the Lower Ninth Ward community. It may not be black and white, but it's there. I thought the idea behind this alliance was that we would bring it all back, but that's not happening. (fieldnotes 2008)

In response to Mr. Holmes comment, another Holy Cross resident organizer, Felicia King, countered:

> There's this misconception that every time that there's resources, that it's predetermined that Holy Cross is gonna get it. There's some folks in this coalition that are very passionate about north of Claiborne [the area of the Lower Nine]. Just because you are passionate about north of Claiborne, it doesn't mean you are dissin' Holy Cross. I know we've been mowing lawns in north of Claiborne for two years; we've been helping people rebuild in north of Claiborne for two years. It is much more expensive to help someone completely rebuild. There are a lot of presuppositions that money is going to Holy Cross. If you think that is an issue, then we need to put it on the table. (fieldnotes 2008)

In this brief exchange we see the social complexity of the Lower Ninth Ward that is often missed in media representations and stigmatized stereotypes of the area. On the one hand, Mr. Holmes insists that some

Holy Cross residents have a history of thinking of themselves as different and even superior when he says, "You talk about segregation, we have segregation within the Lower Ninth Ward community. It may not be black and white, but it's there." Moreover, Mr. Holmes insists that these identity politics (which are matched by quotidian habits of spatial and embodiment production) are still playing out over the recovery process and leading to an uneven reconstruction of the north and south sides of the Lower Ninth Ward. Indeed, in 2010 the northern side of the Lower Ninth Ward was 76 percent less populous than in 2000 and lagged behind the Holy Cross area, which featured a higher rate of resident return (McCarthy 2010).

Some of Mr. Holmes's neighbors, on the other hand, are quick to dismiss his sentiments as "traditional thinking" and a "misconception," but during my six years of ethnographic work in the area, many other Lower Niners have continued to share his sentiment. These feelings were sometimes expressed during neighborhood association meetings in Holy Cross. On one occasion, while discussing two bills being drafted by Louisiana State representative Charmaine Marchand and Senator Ann Duplessis that would grant lay citizens veto power over the City Planning Commission's land use plans that they found objectionable, a resident from the Lower Nine became irate with Holy Crossers who only spoke of the bill's impact on their area. Standing up, the resident raised his voice: "It's like it's always been, one side of St. Claude against the other" (fieldnotes 2009). As I noted previously, Holy Cross organizers usually quickly became defensive when confronted with accusations of elitism. On the same occasion, Mary Hollis, a Holy Cross resident who moved to the neighborhood after Katrina, for example, responded to accusations of elitism from Mr. Jimmy, a community elder who is renowned for his tours of the area. Hollis insisted Mr. Jimmy's "perceptions" were not "reality," but Mr. Jimmy was quick to retort, "But perception *is* reality" (fieldnotes 2009).

Part of the challenge confronted by Lower Ninth residents is that Holy Cross neighborhood organizers often feel charged with intentionally creating social hierarchies and inequities within the neighborhood when, from their perspective, all they are guilty of is trying to improve the area. Nevertheless, the actions of some Holy Cross residents, whether intentional or not, do create socioeconomic differences. The principal neighborhood association in Holy Cross, for example, has a longtime affiliation with New Orleans historic preservation groups that restore and sell homes at values that range between $140,000 and $230,000 and greatly exceed the price of other homes in the neighborhood. By contrast, Jeanell Holmes, a lifelong resident of the Lower Ninth Ward, explained to me how she purchased her historic home for $20,000 and made it inhabitable with a bank loan of $60,000.

Jeanell's shotgun house on Flood Street, although not fully historically accurate in its renovation, cost her a total of $80,000, the maximum sum she and many of her neighbors can afford. Caring for things such as historic preservation, then, enacts subtle forms of exclusion by raising property values, attracting new homeowners who have not resided in the neighborhood, and pushing away more economically humble New Orleanians. Jeanell's example resonates with Gastón Gordillo's (2014) observation that historic preservation groups often err on the side of upholding bourgeois sensibilities concerning the built environment as an unquestionable good. The problem with this perspective, Gordillo, explains, is that it prioritizes elitist feelings of nostalgia over the lives of subaltern populations, thus inciting sentiments of resentment or disinterest toward preservationist agendas on the part of the latter.

In the succeeding section, the social complexity and affective politics (who cares about what and how) of the Lower Ninth Ward have a number of implications for environmentalist projects spearheaded by academics and graduate students from state universities and external non-profits. In Holy Cross, a neighborhood association has developed a separate center devoted to sustainable reconstruction, and this center has become a key node of engagement between academics, external non-profits, and a select group of Holy Cross residents. After Katrina discourses of environmentally sustainable reconstruction have become one of the principal means of envisioning and caring about the area's recovery, but these ways of imagining the area's reconstruction have the potential to be conjoined with those pre-disaster practices many Lower Niners interpret as elitist and as intra-neighborhood segregation. To counter this tendency, some Lower Niners are actively attempting to transform environmentalist discourses, practices, and projects and to invent new ways of being "environmentally friendly" that are sensitive to the social nuances of their two neighborhoods.

Post-Katrina Environmentalism and Sustainability

After Hurricane Katrina, issues of social complexity and spatial production have become key factors in implementing wetland restoration and green housing reconstruction programs organized by academics and non-profits. Although the scientists and project managers involved in these projects have come to the Lower Ninth with the best intentions, some neighborhood organizers have contested the scope and focus of these projects and called for them to adapt to the neighborhood residents' social particularities and self-defined notions of sustainable recovery. These contestations highlight the tendency recognized by Bruno Latour (1993b) of modern epistemology to separate the different dimensions of human experience into distinct objects of scientific inquiry— biology, physics, economics, culture, politics—and these separations do not reflect the intimate ways people, materiality, and meaning are intimately co-constituted in affective ecologies.

Some prominent Lower Nine resident organizers felt academics who spearhead these projects focused on definitions of environment that emphasized nonhuman indicators, such as water salinity levels and cypress forest density when restoring wetlands, and downplayed what the residents cared about most in disaster recovery—the return of displaced neighbors and the restoration of the area's social landscape. Here we see an interesting parallel to Gastón Gordillo's (2014) critique of historic preservation groups, except in this instance, "the environment" (conceived of as everything that is deemed nonhuman or not human-made) is what takes priority over subaltern lives. Meanwhile, graduate students and faculty involved in these projects responded that their programs prioritized community involvement in wetland restoration, but their notions of community involvement figured Lower Ninth residents as mainly supporters of wetland restoration programs and not the focal subject of concern.

These issues surfaced during a meeting called by grassroots organizers in the fall of 2009 that brought together faculty from Louisiana State University, Colorado State University, representatives of affordable and green energy non-profits, and neighborhood residents. The meeting was held at a warehouse that was converted into a gallery and community meeting space by Ward "Mack" McClendon, whom I discussed in the "Introduction." During the meeting, Mack challenged wetland scientists to "not put the wagon before the horse" as he responded to a presentation they made about Bayou Bienvenue's history and restoration. Mack recognized the importance of reforesting nearby wetlands, but he also emphasized the significance of rehabilitating the neighborhood's social landscape, which, in the context of state-supported capitalist disaster recovery, remained an afterthought in city and state reconstruction programs.

> It used to be just like you're talking about, back there. I think if we get that back again, it would be good, and the people—getting back to the people—they totally understand that, given the opportunity, 'specially the elderly. Sixty-five percent of this property was owned by [the] elderly; they're less than five percent back. So I think the key is, don't put the wagon before the horse. (fieldnotes 2009)

In admonishing the landscape architects and wetland ecologists, Mack made the point that the approaches to wetland restoration he had seen thus far featured conceptualizations of sustainable reconstruction that defined the region's ecology in terms that were too narrow. These terms defined the bayou's restoration solely as achieving specific salinity levels and forest density. For Mack and other Lower Nine residents, such conceptualizations of ecology led them to believe that restoring their social lives to what they enjoyed before the storm was effectively made a secondary priority to "environment" as defined by forestry and water science.

In the alternative discourses of environment and ecology articulated by many residents of the Lower Nine, in contrast, the bayou was represented as a gendered space of memories and rites of passage that cannot be considered in the strict categories of wetland science. In these alternative discourses, the bayou is figured as a place of childhood adventure where male residents fished, procured wildlife, and made friendships. On one occasion, for example, Mr. Smith, an elderly Lower Nine resident, recounted his childhood experiences in the bayou. He used the pronoun *we*, indicating that he was always accompanied by other neighborhood boys:

I remember we used to go back on these tracks—we used to call it Florida Walk—and we used to go back there. Sometimes we'd catch a train, go one way, and catch a train and go back the other way. There was a lot of trains because there was a lot of traffic out there on the trains.

But I remember I used to go back there when I was young and go gar fishing. We'd catch garter snakes . . . They had all kinda animals back there. And they had a guy that was further down around Southern Scrap, he had a house out there, and he had a stream that used to go by his house, and he used to rent you boats. And we used to go out and fish and all of that and come back . . .

And some people don't believe this here, I remember they had horses back there, at Southern Scrap! Wild horses! I tell people that, and they don't believe me! They had wild horses back there. Bring you back. This is in the early fifties or middle fifties. (structured interview transcription 2009)

Mr. Smith's comments bring to mind Lefebvre's observation: "If there is production of the city [and the bayou], and social relations in the city [and bayou], it is a production and reproduction of human beings by human beings, rather than a production of objects" (1996, 101). Academic and environmentalist approaches to the bayou's restoration that focus too closely on wetland science and on the bayou as a "scientific object" miss this important insight of critical geography. For residents of the Lower Ninth Ward, the bayou is a place of memories that are intimately linked to the social relations they have experienced over the course of their lives in the area (what Lefebvre would have called perceived space; see chapter 5). These social relations are as much a part of what the bayou and neighborhood *are* as are the former's salinity levels and cypress trees and the latter's historic homes. Thus, the integration of the community into wetland restoration projects must involve more than recruiting the residents' support for environmentalist projects in the Lower Ninth. Instead, returning residents to the area and reinstating the social fabric must be key priorities of environmentalist agendas in this New Orleans neighborhood.

Mack's sentiments were echoed in another context by neighborhood grassroots historian and museum curator Ronald W. Lewis during a meeting between Lower Nine residents and representatives of the nonprofit Make It Right. On this occasion, Make It Right managers arranged a discussion between landscape architects about the geographic integration of the area with the region's wetland features, including the Mississippi River and Lake Pontchartrain. Responding to a landscape architect's comments that focused on the geophysical features of the greater New Orleans area but excluded a discussion of the neighborhood's human landscape, Mr. Lewis commented:

> The people who built this neighborhood worked the sugar cane fields. They bought lots for two hundred dollars and built it themselves. They did it without architects. We didn't have a lot of amenities before. Put our people back, put our tax base back. Without the people, we are nothing. (fieldnotes 2009)

As Mack did, Mr. Lewis insisted that discussions of environmental restoration in the Lower Nine should not sideline concerns about population return. He reminded them that people and their social relationships are the basis of what the neighborhood and its surrounding wetlands *are*, and, consequently, they must be an integral part of what should be restored. As he said, "Without the people, we are nothing." There are different ways one can care about a bayou.

State-Supported Capitalism, Social Complexity, and Reconstruction

When Mayor Ray Nagin ordered the mandatory evacuation of New Orleans in anticipation of Hurricane Katrina's landfall, he created a moment when federal and local government officials, developers, and gentrifying resident constituencies could imagine the city's urban landscape as a space emptied of its pre-Katrina social challenges and opened for "revitalization" through neoliberal practices of disaster reconstruction. These neoliberal practices operated on the assumption that cities were best conceptualized as spaces of capital investment and that reconstruction resources were best used in recovery programs that promised a short-term financial return.

In this vision of urban recovery, the long-term social benefits of public services—for example, public housing, schools, public hospitals—were not considered "cost-effective" returns on governmental investments and were therefore de-prioritized. Examples of these neoliberal policies included demolishing and privatizing more than half of the city's public housing units, creating "target zones" to encourage out-of-state capital investment, closing the city's only major public hospital, and converting the city's public schools to a predominantly charter school system. This disaster reconstruction strategy resulted in the neglect of the Lower Ninth Ward, which had a long history of being seen as a downtrodden neighborhood and therefore not economically profitable. The impact of this approach toward disaster recovery is evident in the slow rate of return of Lower Ninth residents, particularly in the neighborhood's northern section.

City, state, and federal government reconstruction policies based on neoliberal tenets of urban development do not address the underlying relationships between development practice, human values, and material agency that gave Hurricane Katrina its form and magnitude—that is, the lax enforcement of poor environmental regulations, the prioritization of profit-oriented navigation canal construction projects, and the production of racialized class differences through the purposeful structuring of urban spaces. In fact, such policies propose the expansion of the very logics of capitalist cost-benefit that ingrained the catastrophe in all dimensions of social life, including education, housing, and health care. In this context of neoliberalization, non-profits, academics, and activists take it upon themselves to address issues of environmental sustainability and justice. As Kim Fortun (2001) has noted, at the turn of the twenty-first century, post-disaster contexts are moments when populations affected by catastrophes must form new communities with such external actors in order to have a voice in the political and institutional spaces where key recovery decisions are made.

The case of the Lower Ninth Ward gives us a unique glimpse into both the complexities that disaster-affected populations must face when forming the kinds of alliances Fortun had in mind and the importance of approaching these complexities from the vantage point of the ecology of affect. As the ethnographic evidence presented in this chapter shows, disaster-affected localities like the Lower Ninth Ward have complex social landscapes where people differ significantly in terms of their habitus, their racialized class identities, and their feelings about and definitions of sustainable recovery. In the Lower Ninth, we see residents who insist that environmental recovery projects must emphasize the reinstatement of the neighborhood's social landscape by assisting in and ensuring the return of pre-Katrina residents.

The voices of residents featured in this chapter also emphasize the neighborhood's proud history of self-reliance, and they are guarded about the increasing role of non-profit managers and academics in establishing the agenda of disaster recovery. Residents who articulate these positions also express reservations about the social hierarchies and inadvertent elitisms of some established neighborhood associations. In the context of post-Katrina reconstruction, some of these organizations have added environmental sustainability to their pre-disaster focus on historical preservation, but some outspoken residents remain concerned that unresolved space, race, and class-making issues threaten to fold "rebuilding green" into gentrifying tendencies. Nevertheless, these same residents do not necessarily think environmental sustainability is inherently incompatible with their visions of neighborhood recovery; instead, their position is that issues of social justice, embodied in the restoration of the neighborhood's pre-Katrina social landscape, must be central to discussions of sustainable reconstruction.

7. Criollos, Creoles, and the Mobile Taquerias

LATINOPHOBIA IN POST-KATRINA NEW ORLEANS

After Hurricane Katrina, the U.S. federal government—under the direction of President George W. Bush—temporarily suspended the requirement that companies involved in the city's reconstruction obtain documentation from employees and verify their eligibility to work in the United States. The official explanation behind this temporary change in policy was that New Orleanians were entitled to work in the reconstruction of their city and that accommodations should be made for those who lost documents such as drivers' licenses, Social Security cards, or work permits during the storm. Despite this official justification, many New Orleanians interpreted the requirement's suspension as a ploy to make available to reconstruction contractors low-cost and easily exploitable labor—namely, foreign-born workers who lacked a formalized immigration status and were therefore willing to work for lower wages (and in more hazardous conditions) than U.S. citizens and holders of official work visas. President Bush's decision was also considered controversial because many displaced New Orleanians in need of employment would be bypassed for reconstruction jobs as their homes were either flooded or closed indefinitely (the latter applying to public housing residents whose units were closed by order of HANO and HUD) and as they were preoccupied taking care of their displaced families in distant cities like Houston, Atlanta, and Memphis.

Following the catastrophe, the city witnessed the arrival of contractors who specialized in reconstructing areas damaged by tropical storms in the Gulf Coast. Accompanying these companies was an influx of workers from within and without the United States. Studies of this workforce reported that 30 percent of these new arrivals self-identified as Latino, Hispanic, or Latin American, and 25 percent of this subset of laborers did not have a formalized status with the U.S. Immigration and Customs

Enforcement (ICE) service. This latter group of workers reported Mexico, Honduras, Nicaragua, and El Salvador as their nations of origin, although many were already residing in the United States when Hurricane Katrina struck New Orleans (Fletcher et al. 2006; Fussell 2007).

The city was closed for five weeks after its mandatory evacuation in early September 2005. Following its reopening and into the fall–winter seasons of 2005 and 2006, New Orleans saw the proliferation of temporary tent cities as laborers faced the shortage of habitable housing units. Highway underpasses, shopping center parking lots, and the city's parks became campsites for the reconstruction's workforce. But housing was not the only thing in short supply. Lacking kitchens to properly prepare meals and working long hours, laborers also discovered acquiring food was quite a challenge. With many fast food restaurants still damaged or closed and with limited options for cooking at home, reconstruction laborers became a market for ambulatory food vendors, which arrived from out of state to fill this gap. Food trucks specializing in Latin American and Southwestern U.S. fast food offered their customers assorted menus of tacos, enchiladas, and *tortas* (a popular Mexican sandwich served on a bun), and their culinary emphasis earned them the metonymic name "taco trucks." Prior to Katrina, food trucks were rare in New Orleans, but they became ubiquitous during the first few years after the catastrophe. By 2007 Mexican- and Central American–themed mobile food vendors appeared in the major thoroughfares of New Orleans and suburban Metairie. They often set up for business in the parking lots of corporate home improvement stores, where people camped or congregated to be hired as day laborers.

The very visible presence of food trucks occurred at a time when local politicians found themselves pressed to show reconstruction results. Mayor Nagin's second term in office, which he secured through an election campaign that played upon race identity politics in New Orleans, seemed to be developing little traction getting reconstruction projects off the ground. Immediately after the catastrophe, for example, Nagin publicly declared that New Orleans was a "chocolate city," a phrase that conveyed the idea that the city was primarily home to an African American population and that reconstruction efforts should ensure its continuity as such (Grimm 2014). Despite these comments, reconstruction policies under his administration continued both to underserve devastated, historically sociopolitically marginalized, and predominantly African American neighborhoods such as the Lower Ninth Ward and to favor those neighborhoods that were thought to epitomize elite culture in New Orleans, like Uptown (although Uptown is anything but home to a socioeconomically homogenous population).

The mayor's reconstruction policies reflected the political strategies of his first campaign. A former executive for a regional cable company, Nagin had relied on the support of the city's business elite, and his campaign's economic platform catered to these interests, earning him the nickname "Ray Reagan." The mayor, as noted previously, advocated for a recovery plan that was conceptualized within a distinctly neoliberal imagination. Based on a trickle-down reconstruction economics of sorts, the plan would focus on seventeen "target recovery zones" and use public funds to attract out-of-state investors, whose investments in the form of shopping malls and commerce would one day, in an indeterminate future, generate sufficient tax returns to provide the public services needed immediately by disaster-affected New Orleanians. His recovery plan, however, became mired by tax break and back-tax debt scandals among its potential investors, bringing the implementation of the "target zones" to a grinding halt.

With reconstruction off to a stymied start, Mayor Nagin and a small number of city council members seized upon the apparent growth of the city's Latin American and Latino population in an attempt once again to stir identity politics along the lines of race and ethnicity. Nagin, for example, was quick to coin the phrase "No Nuevo Orleans." By replacing the "New" of New Orleans with the Spanish *nuevo*, the mayor conveyed two messages. First, the arriving labor force, with its significant percentage of Latin American–born laborers, posed a threat to an imagined cultural integrity of the city. Second, his administration would not allow this population to make a transformative cultural imprint on the city's social fabric and built environment. Nagin continued to attempt to play on New Orleanians' anxieties about cultural loss in the disaster's aftermath by asking, "How do I ensure New Orleans is not overrun by Mexican workers?" (Pae 2005).

Echoing the mayor's attempt to stir people's apprehensions, At-Large

City Council Member Oliver Thomas also made inflammatory remarks that connected food trucks, the city's imagined cultural identity, and newly arrived laborers: "How are we helping our restaurants that are trying to recover by having more food trucks from Texas open up? How do the tacos help gumbo?" Comparing his out-of-state college experience to that of reconstruction workers, the council member went on to reminisce about his days as a student, saying, "I didn't have a Creole-Cajun food truck. I learned to eat the food. What's wrong with that? What, they don't like our food?" (Krupa 2007a).

Not so far away, in Jefferson Parish, Councilman Louis Congemi also weighed in on the topic. In the suburban city of Metairie, taco trucks set up shop in some of the area's most prominent thoroughfares including Veterans Highway, a multilane road lined with shopping malls and chain restaurants. Congemi advocated for a law that would not outright ban these mobile food vendors but would require them to establish their operations in out-of-sight industrial areas. The councilman explained his position: "Vendors clutter parish streets, they pose safety and health risks and they are mobile, fleeting operations that don't show a commitment to permanent business in Jefferson Parish" (Waller 2007).

There is a significant body of anthropological literature on the relationship between food, taste, memory, and embodiment (Beriss 2012; Farquhar 2002; Sutton 2000, 2001). This literature demonstrates that the tasting body is shaped through practice and in social relations and that food synesthetically evokes memories of people and places. In this chapter, however, I use the controversy surrounding Latin American–themed food trucks in New Orleans to explore the connections between the social production of space, affect (i.e., taste, anxiety, comfort), identity, and racialized difference in New Orleans.

As already discussed in chapters 5 and 6, over the course of the last three centuries in New Orleans, some residents have attempted to create and sustain racialized differences through the structuring of the city's space, delimiting who can live where and what kinds of practices people may (or should) engage in within such space. My use of "attempted" is important here, as this effort has always been more of a tendency than an accomplished task. At the same time, the history of New Orleans is also one of migration and culture change. Newly arrived settlers came to live in proximity with diverse cultural groups and found themselves borrowing and developing new quotidian practices and traditions, just as their host communities did the same, driving a process of cultural exchange and genesis that is sometimes recognized as creolization. The history of New Orleans is the story of populations who were at one point perceived as being a cultural other but whose mimetically evolving material, culinary, and architectural culture eventually became iconic of the city itself.

Rather than being defined by a fixed culture that can be threatened or overrun by an immigrant horde, New Orleans is more an ever-emergent space of mimesis and creolization where some residents attempt to fix racialized differences and achieve as much success at creating a rigid racialized order as much as they fail. In the mid-nineteenth century, for example, established francophone and Anglo-American New Orleanians perceived populations immigrating from Ireland, Germany, and southern and eastern Europe as culturally distanced others. These new arrivals made space for themselves and established neighborhoods such as St. Roch, Marigny, and Irish Channel. As immigrant groups attempted to settle in the city, they were subjected to discrimination and violence. In 1877, for example, twelve Italian-born New Orleanians were lynched after one of their compatriots assassinated the city's sheriff (Campanella 2006). Today the urban spaces these new arrivals created are considered indispensable parts of the city's architectural landscape, and one example of immigrant community culinary culture is the muffuletta sandwich (a food whose American development is credited to Italian immigrants). This food is now considered a quintessential New Orleans dish.

The descendants of mid-nineteenth-century immigrants who were at one time perceived as culturally foreign others today make unchallenged claims to being New Orleanians and engage in practices such as consuming po'boy sandwiches and organizing Mardi Gras–like ethnic pride parades that celebrate their hyphenated identities as Irish-, German-, and Italian-American New Orleanians. But what is the process that an immigrant population on the path from "other" to New Orleanian experiences? Is the process the same for all immigrant populations, or do members of some groups receive differential treatment depending on circulating narratives about the nature of their difference (e.g., racial and insurmountable versus cultural and behavioral and yet capable of being reduced), in what Ruth Mandel (2008) has called the differentiation of difference? What are the nuances of such a process? How will these newly arrived workers change (or creolize), and how will they themselves influence the continued (re)creation of the city's urban landscape? And finally, what are the affective dimensions of this process? How does a behavior or a person's appearance shift from being something that evokes anxiety (about culture loss, about the perceived danger of the other) to something that creates a sense of familiarity?

Criollos and Creoles

In this chapter I explore the concept of creolization as a tool for thinking about difference, change, and cultural emergence. As I write this chapter in 2014, I find this exploration more and more relevant as news stories increasingly discuss the cultural impact that immigrants from Latin America will have on the United States. These discussions are often framed in terms of assimilation, or the idea that an immigrant population will undergo a process of culture change, blending into its host society and having little impact on it. Take, for example, the recent case of Jason Richwine. The Harvard graduate in public policy and Heritage Foundation analyst gained national attention for claiming that "Latinos" cannot assimilate to the U.S. mainstream because they have a different ancestry than early twentieth-century immigrants to the United States and are thus more culturally removed from Anglo-American "culture" (Carrasquillo 2013).

The concept of assimilation resonates with a metaphor commonly used in the United States to speak of the cultural dimensions of transnationalism—the melting pot. The figure of the melting pot suggests that immigrants will become a part of a larger host society and lose their distinction in the process. Although this metaphor does allow some room for imagining the contribution of new arrivals to the greater whole (the melted substance, after all, contributes to the larger aggregate), that whole is still envisioned as a homogeneous mixture (Basch, Schiller, and Blanc 1993). As debates over immigration flare in the United States, I find both concepts of assimilation and the melting pot to fall short of grasping the processes at play in the flows of people, material culture, and body-cultivating practices that have characterized the inter-connected life of the United States and Latin America for more than a century now. Both concepts, for example, assume that the host society is a homogenous group. In this way, these notions of culture change fail to recognize the diversity of the United States in terms of class, regional, and ethnic differentiation. Furthermore, they make simplistic assumptions about processes of culture change and ignore the mimetic, creative, and embodiment-shaping practices involved.

As explored in chapters 4 and 5, the concept of creolization originates from the dilemmas concerning identity, practice, embodiment, and ontology (what something or someone *is*) that were engendered during European colonial expansion from the sixteenth to the eighteenth century. In using the word *criollo* Iberian-born Spaniards (and, later on, other continental Europeans) were attempting to apprehend processes of culture change brought about by social intimacy between settler and colonized populations (Cañizares-Esguerra 2002; Stoler 1995). The challenge that emerged for colonial settlers was how to maintain their distinction from local populations, given the propensity of people to engage in mimetic acts (Taussig 1993). Mimesis, in this case, is the human tendency to imitate another, or what is colloquially referred to in U.S. English vernacular as a person's "rubbing off on you." Mimesis can take place in a variety of ways for a number of reasons. In some instances, it is an unconscious imitation of bodily gestures, stances, and linguistic patterns of those who are proximate. In other instances, it is a strategic attempt to copy the behaviors, ritual practices, and culinary traditions of someone who is deemed other in order to capture their fetishistic, hegemonic, or suggestive cultural power (Langford 1999; Taussig 1993). Scholars of mimesis have gone so far as to use the term to denote a person's reaction to another, or an action that does not so much involve imitation as a gesture in relation to someone else (Gibbs 2010).

Because mimesis is a bodily practice, it is not necessarily an exact imitation but an interpretation of another's bodily movement and gestures from one's own embodied disposition (the way a person is in, experiences, and moves his or her body) and, in the process, creates a new cultural expression. Mimesis, then, runs counter to the idea of assimilation; rather than a vanishing of difference, it is a proliferation of novel forms of being. It is also worth keeping in mind that mimesis takes place within broader cultural contexts that often involve ethnocentric hierarchizations. Thus, imitation—whether conscious, strategic, or unintended—is always subject to interpretation and value judgment. It incites the question, what does it mean to act like the other?

In New Orleans, as in Dutch Java and New Spain, the term *Creole* was initially used to denote descendants of French settlers who, when exposed to the cultural practices of African diaspora populations, Native Americans, and other European settlers, devised novel quotidian practices of the care and formation of the self, giving rise to different ways of inhabiting the body (Dawdy 2008; Hirsch and Logsdon 1992). In all of these colonial contexts, continental observers did not interpret the new embodied cultural expressions in a positive light, and they used the term *Creole* to denote a category of lesser European.

Using creolization as a heuristic device, however, is not without complication. Tom Boellstorff (2005), for example, has argued that the concept assumes the existence of racial and cultural wholes or purities, as it is used to denote the joining of two distinct things that then produce a third culture or embodiment. While this is an important critique, it is worth noting that other scholars such as Shannon Lee Dawdy (2008) have made the case that while European colonizers may have made claims to cultural and racial purity, that is all they were—claims. Creolization is a powerful thinking tool, Dawdy insists, because all people are in one way or another creolized, even if they claim purity just as self-identifying Creoles like Oliver Thomas did in post-Katrina New Orleans. In our use of creolization, then, we must be careful to differentiate between people's claims to purity and wholeness and the complex mimetic processes involved in their embodiment of cultural practice.

My use of creolization in this chapter is not meant to reiterate the ethnocentric colonial hierarchies and claims to purity behind the original use of the term *criollo*. What I propose, instead, is that criollos, Creoles, and creolization are "good to think" when reflecting on the mimetic and culturally creative processes that are bound to happen when people with different life experiences, customs, ways of experiencing affect, and embodiments come together in post-disaster contexts. Thinking in terms of creolization shifts the discussion about the arrival of new Latino populations away from concerns about culture loss to one of cultural proliferation. Such a shift encourages us to change Mayor Nagin's rhetorical question from "how do I ensure New Orleans is not overrun by Mexican workers?" to "what innovative ways of being New Orleanian will newly arrived reconstruction workers generate, and in what way will they write the latest chapter of the city's history of migration, creolization, and ongoing emergence?"

Situating an Ethnographer's Affect

In the summer of 2006, when I was beginning my journey as an anthropologist of post-Katrina reconstruction, I sat down to eat lunch and talk about my research plans—namely, the ethnography of recovery planning—with another anthropologist who lives and works in New Orleans. My colleague listened to me patiently and, when I was done sharing my ideas, proposed that I focus on a different topic. "You should study the Latinos who are coming to work on the reconstruction," I remember the person saying. "You are uniquely positioned to do this research." The suggestion bothered me, and exploring the reason I experienced this affective reaction may help unpack some of the epistemic, colonial, and historical baggage wrapped up in the ethnographic project that concerns this chapter.

In my own history of transnational migration, I developed an aversion to the terms *Latino* and *Hispanic* as a means of self-identification. During my early life experiences, I had internalized other categories that situated me in terms of national, gender, class, and ethnic identity. I thought of myself as an urban male Ladino, as from zone 11 in Guatemala City, and as a member of a family with tenuous claims to middle-class status and with aspirations of upward social mobility. These ways of self-identifying were by no means unproblematic; they were mobilized in Guatemalan nation-building practices that resulted in deplorable acts of everyday ethnocidal discrimination and genocidal violence against indigenous populations, rural "peasants," and Ladino labor and leftist organizers.

Without question, my migration to the suburbs of New Orleans as an adolescent played a critical role in the development of my anthropological sensibility. My experience of being perceived—and being differentially treated—as Hispanic or Latino by classmates, teachers, and neighbors made me reflect on the ways power operates through language and how terms such as *Ladino*, *Indio*, or *Latino* attempt to homogenize and racialize human populations and fix their position in a society's body politic. I felt the terms *Latino* and *Hispanic* erased the diverse histories of people who migrate (or whose ancestors migrated) from Latin America to the United States, lumping together individuals with varied attachments to places and quotidian embodiment and identity-forming practices.

The terms *Latino* and *Hispanic* also seemed to draw impermeable cultural boundaries between Latinos and non-Latinos, defining a difference that was absolute. Such a way of imagining difference seemed to deny the cosmopolitanism of many Latin American immigrants. As a young Guatemalan, my life experiences included feasting on tamales during holidays and frequenting religious parades during Holy Week as much as they involved consuming North American popular culture, frequenting expatriate French pastry shops in Guatemala City, and taking family vacation trips to New York and New Orleans. Over the course of my life history, my body became one that experienced pleasure in relation to foods, flavors, and spaces with diverse genealogies, making things such as the pungent smell of banana leaf–wrapped tamales and the suburban streets and spaces of Guatemala City evoke sentiments of comfort, familiarity, and nostalgia just as the theme songs of blockbuster Hollywood films do.

A careful reader may take objection to my critical discussion of the terms *Latino* and *Hispanic* and argue that I am conflating categories used to describe U.S.-born people of Latin American or Iberian ancestry with first-generation Latin American immigrants. In this chapter I emphasize that, in practice, the term *Latino* is often used colloquially in the United States (and by the U.S. Census Bureau as well) to refer collectively to people with these varied backgrounds and that Latin American immigrants also often choose to self-identify as Latino. The categorical boundaries of these terms, then, are regulatory ideals whose lines become blurred in everyday usage.

Scholars who work on issues of identity, embodiment, cosmopolitanism, and globalization, such as Arjun Appadurai (1996), Homi Bhabha (1994), and Elizabeth Povinelli (2002), recognize concerns about categories as indicative of the fundamental double binds of identity and identity politics in our postcolonial world. Appadurai (1996), for one, has made the case that the flows of capital, commodities, imaginings of development, and people across national borders in the post–Cold War era have given rise to ways of being, of thinking about the self, and of identifying that trouble existing definitions of "culture" and "nation" as neatly bounded and monistic entities. In a similar line of thought, Bhabha (1994) has noted that groups unified through identity claims always comprise people with heterogeneous embodiments, life histories, and, I would add, ways of experiencing affect.

At the same time, the identity politics surrounding categories such as Latino can provide people who experience discrimination and marginalization with an important mechanism for building constituencies and challenging institutional and everyday discriminatory practices (Hill and Wilson 2003). For Elizabeth Povinelli (2002), in turn, ethnic recognition involves cunning, it is not a given, and it requires sociopolitically marginalized people to be savvy about the ways they navigate the categories that state governments use to document, think, and speak about difference. Consequently, I do not argue that one way of identifying (as a Latino or as a Guatemalan, middle-class, male Ladino) is more adequate than another but that power operates through all identity categories. Rather than an objective description, the act of stating who someone *is* becomes an ontological claim and a political move to be subjected to interpretation and analysis. Anthropology's task is not to make prescriptions on what terms people should use for self-identification but to provide us with the tools to think about the pitfalls and possibilities of identity claims.

But returning to that post-Katrina conversation with my senior colleague, some of my resistance also stemmed from a refusal on my part to be relegated to a particular topic because of my ascribed status as a Latino. Certainly my Spanish language skills would facilitate my communication with this population, but I was not fully convinced I would have ready access to this diverse community. In my previous ethnographic experiences in post-Mitch Honduras, for example, Limón de la Cerca residents often questioned my claims to "Latinoness." I distinctly recall one occasion when, while conducting household surveys with

Rosa Palencia, my Honduran interlocutor, a Mara Dieciocho gang member called her aside and asked her, "Where is that gringo from?" Rosa later jokingly retold this event and pointed out that my diasporic life had shifted my bodily comportment in such a way that no one recognized it as Central American any longer.

On other occasions, I had lengthy discussions with Cholutecans about local race categories during which they claimed I was *chele* (fair) even as I insisted I was *trigueño* (of dark complexion). What these observations about my appearance indicate is that people are by no means objective observers of physiological traits. Instead, they interpret bodies through senses that are tacitly culturally trained. What is more, in the act of seeing difference, people not only observe skin color and physiological traits but also interpret the subtleties of bodily movement, stances, and the commodities a person uses to index social positioning and identity.

Southern Hondurans were right in not recognizing me as one of their own. I was a creolized Guatemalan Ladino who had come of age in the suburbs of New Orleans and matured as an anthropologist in the graduate program at the University of Florida, and this history was inscribed in my body. Suggestions that doing research among Latinos would come easily to me, then, seemed to make assumptions about homogenous forms of embodiment, behavior, and identity that I felt did not grasp the diversity of people who either self-identify or are identified as Latino.

But if I had these reservations about documenting the experiences of Latin American and Latino laborers in New Orleans, how did this chapter become possible? Given the importance of socially produced space in the making of New Orleans's built environment and its residents' racialized identities, as outlined in chapters 5 and 6, the stories in local newspapers about the proliferation of so-called taco trucks and the threat these vendors supposedly posed to the city's cultural integrity seemed to speak directly to my developing interests in New Orleans. If neighborhood and suburban spaces were spatial effects of attempts to make racialized differences and bodies, then the taco trucks seemed to subvert the city's spatial order. In the suburban context of Jefferson Parish, their ease of movement into areas such as Veterans Highway—whose imagined suburban space of "white flight" required painstaking labor on the part of police departments (via racial profiling practices) and school systems (through de facto forms of segregation)—seemed to arouse xenophobic and racially charged anxiety among politicians and some residents. In the suburbs and in New Orleans, I saw the occasional handwritten sign attached to light posts echoing Mayor Nagin's words: No Nuevo Orleans.

In the more central urban areas of Orleans Parish, furthermore, the mayor and city council members represented mobile food vendors as a threat to the city's Creole culture. Once again, the trucks' mobility, combined with their overt Latin American food theming, seemed to evoke apprehension, as if they were capable of upsetting the city's already disaster-destabilized socio-spatial order. But what actually happened? Did newly arrived Latin America–born and Latino laborers really overrun the city and transform its built and social landscape?

Latino Ethnicity by the Numbers in Pre- and Post-Katrina New Orleans

What was missed in the statements made by Mayor Nagin and Councilmen Oliver Thomas and Louis Congemi was that immigrants born in Latin America and their descendants were already present in New Orleans and were part of the city's urban fabric long before Katrina. Furthermore, even with the influx of newly arrived reconstruction workers, the percentage of people who could be identified using the term *Hispanic* (whether born in the United States or not) out of the total population of Orleans Parish and the greater New Orleans fell below the national average for other U.S. urban areas before and after the disaster.

In the 2000 U.S. census, 32,418 people self-identified as Hispanic in Jefferson Parish, and 14,826 did so in Orleans. In the 2011 census, these numbers went up to 54,815 and 18,927, respectively. In percentage terms, the census figures suggest the Hispanic share of the total population grew from 7.1 percent to 12.7 percent in Jefferson and from 3.1 percent to 5.2 percent in Orleans Parish. By comparison, the average percentage share of people who self-identified as Hispanic while living in other U.S. urban areas in 2011 was 16.9 percent, meaning the Hispanic population of both New Orleans and its western suburbs remained well below the national average (Mack and Ortiz 2013).

The census numbers, then, suggest that the city was far from becoming Nuevo Orleans. But these numbers tell only a limited story about identity and culture in New Orleans, leaving one to ask: Who were the people who self-identified as Hispanic in the 2000 and 2011 census data? How did these people think about their identities outside of the specific categories of the U.S. census? What kinds of quotidian disposition- and affect-shaping practices did they engage in? What roles did they play in shaping the landscape of New Orleans, and in which ways did life in the city's broader environment shape their experience of the senses, sentiments, and the self?

Nuances of Latino and Hispanic Ethnicity before Katrina

New Orleans has a long history of being home to people whose first language is Spanish or who immigrated (or are the descendants of people who immigrated) from the Iberian Peninsula, the Canary Islands, and various parts of Latin America. In 1763 at the end of the French and Indian War, France ceded New Orleans and parts of Louisiana to Spain for a forty-year period. Spanish presence in the area, however, was not significant until 1766. One of the most notable migration processes in the late 1770s was the settlement of areas in the vicinity of New Orleans such as Barataria and St. Bernard Parish with Spanish-speaking people from the Canary Islands. Their descendants today are recognized as Isleños, or "Islanders."

While the cultural impact of the period of Spanish control is sometimes thought to be minimal (see Campanella 2006), Isleños and Iberian settlers did undergo a process of creolization and became part of the contemporary landscape of the greater New Orleans area. Today the descendants are considered part of what New Orleans and southeastern Louisiana *are*. More than 230 years later, surnames of Spanish origin are not unheard of in New Orleans and its outlying areas, and the people's quotidian practices, linguistic patterns, and physiological appearance are no longer considered foreign to the region.

Significant migration of people from Latin America, however, did not occur until the early twentieth century, and it is often associated with the close economic ties that developed between the city and Central America, particularly Honduras and its Bay Islands. In 1899 Joseph,

Luca, and Felix Vaccaro, owners of Vaccaro Brothers Corporation, began importing fruit—principally bananas—from Honduras. The Vaccaro brothers were the children of Italian immigrants Stephano and Maria Vaccaro, who migrated to the United States in 1860 and established themselves as fruit and produce merchants. The Vaccaro Brothers Corporation went on to become Standard Fruit, which is today's Dole Food Company (Leonard 2012). A similar New Orleans–based enterprise was Cuyamel Fruit (which went on to become United Fruit Company and is today's Chiquita Brands International). It was founded in 1911 by Samuel Zemurray, a Russian immigrant who began his career as a fruit commerce entrepreneur in Mobile, Alabama, and eventually moved his operation to the city of New Orleans (Leonard 2012). Advances in steamboat technology and refrigeration allowed the owners of both companies to develop business strategies that involved importing fruit from tropical regions, namely Honduras's Caribbean coast. The companies' operations resulted in a growing transnational movement of people and commodities between New Orleans and northern Honduras, and Central American–born populations went on to make their mark on the city, creating an enclave along Banks Street in the Mid-City area.

As discussed in chapter 2, the development of these transnational trade circuits had significant implications for political culture, labor, and land tenure practices in Honduras. But the development of the fruit companies also had impacts—arguably much more favorable ones—on New Orleans. In 1960 historian Edwin Adams Davis wrote of the companies' socioeconomic effects on the city:

> What the United Fruit Company has done in the way of developing banana-growing and other industries in Middle America has been of inestimable value to New Orleans, for during the past half-century it has opened up trade channels not only for its own products, but has supplied the facilities such as transportation, both water and rail, for others to develop industries and trade. Moreover, acting as a goodwill agent, it has brought about a better understanding between North Americans and Latin Americans, who, speaking a different language and living according to different customs, were strangers in business and in their social life, although they did not live far apart. Thus, the

United Fruit Company is essentially a part of the history of the City of New Orleans. (Davis 1960, 31)

Over the course of the twentieth century, U.S. foreign policy continued to tie the social landscapes of Central America and New Orleans. When the United States conducted political and military interventions in Guatemala and Honduras to protect the fruit companies' interests and became involved in Cold War conflicts in El Salvador and Nicaragua, a new trajectory of social polarization, repression, and counterinsurgency began in Central America. It destabilized most of the regions' nation-states, driving millions to migrate to Mexico and the United States.

By the late 1980s New Orleans and its growing suburban areas counted a noticeable presence of Nicaraguan-, Honduran-, and Salvadoran-born residents. Kenner and other suburbs became known as strongholds of Salvadoran immigrants, while areas such as Metairie's Fat City, with its low-cost apartment complexes, also attracted Central Americans of more modest means, my family included. How did these "new New Orleanians" think about themselves in this novel setting? If people's embodiments and affective dispositions are the product of life experiences in socially produced spaces, what kinds of experiences were these Central Americans having in the context of southeastern Louisiana, and what kinds of agency were they manifesting in the quotidian making of the city?

In the summer of 2014, I sat down with Carolina Hernandez to talk about these questions. Carolina is a U.S.-born descendant of Honduran parents who moved to New Orleans in the mid-twentieth century. Today she is employed by one of two non-profit organizations in the New Orleans area that help Latin American–born immigrants make their transition to life in the United States. I began our conversation by sharing the idea for this chapter, leading Carolina to consider the politics of identity among Central Americans living in the city before Katrina. Carolina commented:

> I find that generations of Latinos who were born here, like myself, many don't identify as Latino. We are not a Latino-centered city. It's hard to identify like that when you don't have a group that you look at and say, "Oh, that's a group that I belong to."
>
> I was just getting my nails done by a woman who's been here since

the eighties, for example, and she was impressed that I could speak Spanish. She said, "Most people who I know don't speak Spanish so well." My mom has told me this story about when she went to enroll my older sister in school, and she did not speak English. The teacher fussed at my mom. She told her she should be teaching her English, not Spanish. When she went to enroll me, they fussed at her again and said she was doing her daughters a disservice. Many Latinos like my parents who arrived [in the] early sixties, they did not teach their children Spanish.

There is a base of Latinos like myself, my generation and older, who have always struggled with this identity of Latino. There wasn't a sense of camaraderie, that we are Latinos; there is no sense of place. I only resolved that for myself by seeking out my identity for myself. There's another person here in the office who has a similar reflection on that, and he is struggling with his identity. (structured interview 2014)

Carolina's narrative shows the subtleties of differentiation among people who could claim a Latino identity in pre-Katrina New Orleans. Different life experiences such as being born in the city and chastised for deferring English-language education or migrating as a fluent Spanish-speaking adult from Latin America become important distinctions among Latin American–born immigrants and their descendants in her recollection. Most important, for Carolina, Latino is not a "naturalized" or spontaneous form of identification, so it is not a given that she will think of herself as a Latina. Instead, she feels that such an identification is contingent on socio-spatial factors—for example, on whether an existing and visible group of people feels the need to identify as such and creates places (social and material) that make this identity tangible. As she said, "New Orleans was not a Latino-centered city."

In the absence of a Latino "sense of place," Carolina engaged in quotidian and mimetic practices that creolized her as a New Orleanian of Latin American ancestry but not necessarily as a Latina. This process of creolization involved developing an affinity for spaces of socialization, foods, and musical styles that, although Latin American influenced, were thought to index a New Orleanianness more than a Latino identity. As we spoke about this idea, she offered examples of music she would listen to and nightclubs she frequented prior to Katrina—Café Brasil and Cafe Istanbul on Frenchmen Street, a well-known nightlife area. These clubs featured Latin American music and dancing but catered to a diverse audience. Rather than being seen primarily as "Latino" clubs, though, they were regarded as a symbol of the city's Caribbean cosmopolitan character.

With the arrival of Latin American and Latino workers after Katrina, Carolina and other New Orleanians of Latin American ancestry had to negotiate their identities and affective dispositions with this group of new New Orleanians. She went on to explain:

> You have this growing base of Latinos who consider themselves New Orleanians. Then you have this group of Latinos coming in from other parts of the country, strongly and overtly identifying themselves as Latinos, so they experienced a sense of rejection from other Latinos [established New Orleanians]. "Why this discomfort with these new Latinos?" I asked myself. We never brought attention to ourselves in a significant way, and now you have all these Latinos who are not from here calling attention to themselves, and I kinda love that about them. My only concern with these Latinos is that sense of "they know who they are and don't apologize for it." In some sense, because they maybe felt that rejection, they may not want to participate in the culture of New Orleans, so they reject the idea of assimilating. They would never do that. That's probably a survival thing: "I need to be true to who I am, otherwise I lose myself." How do you prosper and not lose a sense of who you are and integrate yourself within the culture?
>
> You can drive down any street right now in Mid-City, and they'll put their music on, and they don't make much of an attempt to reach out to their non-Latino neighbors. They have been targeted and harassed. We have a group of Latinos who we work with, and we get stories that shock us about what they have experienced. Even though they have experienced crime, they still want to stay; they don't want to leave. For me, that's a really strong indicator that there's hope that we can find a way for them to find their self-identity and find their cultural traditions and integrate into the community they are living in. (structured interview 2014)

I find it interesting that Carolina chooses to speak about her experiences negotiating the politics of identity with newly arrived Latinos in affective terms. In her narrative, she asks herself, "Why this discomfort with these new Latinos?" Her question reveals that alterity, or otherness, is sensed before it becomes a concern of consciousness. But the body that feels is also a body whose affective responses are cultivated over the course of life experiences in a social environment that has its unique history of differentiation, be it in terms of class, ethnicity, gender, or race.

Carolina's theorizing of her affective experience of difference with other Latinos connects her bodily reactions, like discomfort, with the city's body politic. Specifically, she makes a connection between her comfort and both the space that descendants of Latin American immigrants hold in the social landscape of New Orleans and the ways newly arrived Latinos challenge—and perhaps destabilize—this state of affairs through their identity claims and use of urban space. It is telling that Carolina's words and experiences resonate with those of the Tremé residents discussed in chapter 5; their experiences of frustration with disaster recovery-planning processes were also contingent on their position in the city's social order.

Carolina framed her discussion of identity and practice among Latinos in post-Katrina New Orleans in terms of assimilation ("My only concern with these Latinos is that . . . they reject the idea of assimilating"), leading us to consider the ways we think about difference, culture, and change in North America. The concept of assimilation suggests that newly arrived populations should blend into an existing cultural landscape and become part of the host community. In this way, the other becomes like the self, supposedly inhabiting the same kinds of embodiment and subjectivity as the established members of a given community. By the same token, a refusal to assimilate assumes that populations such as the newly arrived Latinos remain culturally unchanged. But what if assimilation, with all its inherent assumptions, falls short of grasping the mimesis and creolization at play in places like New Orleans?

The case I am making here is that, rather than blending new arrivals into host communities, the processes of creolization involve transforming all involved parties whether it be through mimesis, through the creation of new social and material spaces, or through novel expressions of culinary culture. Moreover, the mimetic practices involved in creolization are not always conscious or overtly strategic, although they can reflect hegemonic notions of what or who is desirable or laden with fetishized power. Mimesis is something people *do*, sometimes consciously and with a specific intention, sometimes inadvertently although nevertheless in culturally and power-laden ways. Carolina's narrative begs the questions: Even if newly arrived Latinos resist assimilation, will life in the greater context of New Orleans shift their quotidian embodiment-shaping practices and affective attachments? And in what ways will these Latinos also shape the ever-emergent urban landscape of New Orleans?

SPACES, PRACTICES, AND EMBODIMENTS

In the years since Katrina, new Latino arrivals have not overwhelmed the city and radically transformed its cultural tone into a Nuevo Orleans. New Orleans has certainly shifted demographically, but the biggest threat to its imagined cultural integrity actually came from a wave of progressive, highly educated, relatively young, and childless people who arrived to work in the non-profit and private sectors during the reconstruction (Mack and Ortiz 2013). When combined with the loss of more than 110,000 self-identified African American residents because of the prolonged closures of public housing, the inefficiency of the Road Home Program, the layoffs of public schoolteachers when the school system privatized into charter schools, and the obstacles in securing insurance settlements for damaged housing (Adams 2013), the city has seen a change, but it cannot be solely attributed to an influx of Latin American and Latino reconstruction workers.

Still newly arrived Latinos have had an impact on the city. The most visible of these impacts has come in the form of restaurants and clubs that cater specifically to these new New Orleanians. Before Katrina, the city's most celebrated Mexican-themed restaurant was the Salvadoran-owned Taqueria Corona, located in the Warehouse District, an area renowned for its revamped warehouses turned into costly condominiums and art galleries. Post-Katrina, a number of high-visibility establishments appeared in major thoroughfares like Carrollton and Tulane Avenues. Businesses such as El Rinconcito (a bar catering to predominantly male clientele) and Taqueria Guerrero (a restaurant that is now lauded by established residents and new arrivals alike as the only place to find real Mexican food in New Orleans) opened next to the ice cream and pastry vendor Angelo Brocato. Founded in the early 1900s by Italian immigrants who were once considered other and foreign, Angelo Brocato is today valued by many New Orleanians as an indispensable part of the city's culinary culture and urban landscape.

Beyond commercial establishments, newly arrived Latinos have also clustered in their residential patterns in the Mid-City area of New Orleans. The New Orleans Recreation Development Commission worked closely with Latino advocacy organizations in developing one of the area's parks, Easton, as a welcoming space for this population. In the summer of 2014, for example, flyers advertising the Latino Heritage Festival at Easton Park encouraged potential attendants to "represent your Latino country" and "celebrate your Latino heritage in New Orleans."

Reactions on the part of established New Orleanians to newly arrived Latinos have been mixed, ranging from blatant acts of hostility to attempts to bridge perceived cultural gaps. In Tremé, for example, Cheryl Austin once commented during an afternoon stroll that recently arrived Latinos and Latin Americans (who now had a visible presence in the area) engaged in quotidian spatial practices she found similar to those of subaltern African Americans—for example, hanging out and using sidewalk and porch spaces to socialize. "Your people are just like my people," I remember her saying. I interpreted her words as an attempt on her part to bridge what she perceived as potential cultural differences among her neighbors, new and old, and ourselves.

Not all New Orleanians were as welcoming as Cheryl was, however. Through Carolina Hernandez's organization, city administrators invited the children of Latino and Latin American immigrants to paint a mural on one of Easton Park's structures. The painted mural included a text written in Spanish, but established neighborhood residents defaced it. While attempting to restore the mural, Carolina reported being approached by a neighborhood resident who insisted that Spanish was not a native language of New Orleans and had no place on such a prominent neighborhood landmark.

The spaces that newly arrived Latinos are creating are not solely physical and commercial; they are also juridical. Reconstruction contractors in the aftermath of Katrina eagerly exploited Latino labor. Reports abounded of Latino laborers not being paid for their work and being exposed to hazardous working conditions and materials without appropriate protective equipment (Fletcher et al. 2006). These laborers also experienced persecution by U.S. Immigration and Customs Enforcement, which often works with local law enforcement agencies to conduct raids, arrest foreign-born workers who do not have U.S. work visas, and often deport them to their countries of birth. These conditions led Latino advocates in New Orleans to establish a chapter of the Congress of Day Laborers, a national organization that advocates for the rights of immigrant workers.

Through its organizing practices, the congress successfully challenged ICE–New Orleans Police Department collaboration. Using organized protests at ICE offices in downtown New Orleans and carefully calculated acts of civil disobedience, such as blocking major streets in 2013, the congress and its members called attention to their contribution to the city's reconstruction and the contradictions of their treatment as exploitable labor and undocumented immigrants. Following the Congress of Day Laborers' mobilization, the police ceased collaborating with ICE in conducting raids, making New Orleans more hospitable to Latin American–born residents who lacked a formalized immigration status.

Although difficult to count and estimate, community organizers in New Orleans are certain that the post-Katrina Latino population continues to grow as families in other states such as Arizona and Georgia hear of the more favorable conditions in this part of Louisiana. Nine years after the storm, Latinos were still making their mark on the city by creating various forms of spaces, but they are by no means culturally overwhelming the city. In this way, they have written a new chapter in the long history of migration, adaptation, and creolization of the cityscape. But in what ways are Latinos themselves changing?

Fernando and Katherine, a Post-Disaster Love Story

When reflecting on the cultural impact of life in New Orleans on the new Latino residents, Carolina Hernandez felt this population was resisting assimilation and that, unlike established pre-Katrina descendants of Latin American immigrants like herself, they were hesitant to partake in the city's broader culture. With the creation of recreational, social, juridical, and residential spaces discussed in this chapter, it was certainly apparent in 2014 that this population was making strides to (re)create a distinct identity, form of embodiment, and affective disposition that differentiated it from other established New Orleanians. Carolina, for example, remarked on how new Latino arrivals use music to cultivate affective preferences when they play *ranchera* on radios in street settings. This Mexican and Southwestern U.S. genre, in her opinion, differed significantly from the musical preferences of established Latinos before Katrina—for example, Caribbean-influenced salsa music. Speaking about this distinction, Carolina switched from English to Spanish for emphasis, saying:

> *Tienen su propia cosa* [They have their own thing]. They have a different way. It's very hard for me to make sense of it because I don't relate. They have their *ranchera*, they have what they like. (structured interview 2014)

But creolization can occur in multiple ways, leading either to a sense of shared identity as a creolized New Orleanian or to new ways of manifesting a Latino identity. In this case, even if newly arrived Latinos *tienen su propia cosa*, life in the broader context of New Orleans will certainly have an impact on their daily and ritual practices, leading to a novel expression of *Latinidad*. Moreover, processes of creolization are multigenerational. The term, after all, was devised to speak about the descendants of colonial settlers, not the initial migrant generations.

Still there are cases that may shed some light on how the dispositions of some newly arrived Latinos will shift over the course of their lives in New Orleans. Katherine Crane, for example, is a New Orleanian of Scottish, Irish, and English ancestry. Some of her forebears arrived in the city at the end of the eighteenth century and settled in Tremé. Over several generations her family creolized in such a way that one could argue it is now quintessentially New Orleanian. Her uncles are members of established Mardi Gras krewes (members-only clubs that organize Mardi Gras balls, floats, and parades), her grandparents were owners of a prominent local clothing store, and she grew up in the affluent Uptown area of the city. Today Katherine and her extended family are active participants in New Orleans's public culture, whether it is watching parades from her grandmother's balcony overlooking St. Charles Avenue, one of the city's major thoroughfares, or attending Saints football tailgate parties.

Fernando García, meanwhile, was born in Michoacán, Mexico, and migrated to California as a young adult in search of fair wages and opportunities for economic advancement. Before Katrina, he traveled throughout the Southwest with his father and brother, working manual labor jobs in various cities. When Katrina struck, the Garcías moved to New Orleans to work on the recovery.

Since 2006 Fernando has worked a variety of jobs, from shoveling mud and debris out of homes, to light construction, to working the door at bars in the French Quarter. Fernando and Katherine met at a club where he worked, started dating, and eventually moved in together. Two years later, Fernando and Katherine's Facebook pages were filled with photographs of them partaking in the city's public life. One photograph shows Fernando wearing Saints football paraphernalia outside the Superdome (the city's major stadium); another shows him with green face paint at a St. Patrick's Day parade.

In July 2014 I was invited to Katherine and Fernando's home in Mid-City, where they organized a barbecue to entertain Katherine's multiple cousins, siblings, and in-laws. At the gathering, Fernando mingled with his new kin group, most of whom have significantly different life histories from his. Eventually, we made time to talk. Fernando had watched the Mexican team play in the World Cup earlier in the day, and the team had lost. He watched the game at Finn McCool's, an Irish-themed pub in Mid-City that caters to university students and young professionals. I asked Fernando how he thought he had changed over the course of his life in New Orleans. After a pause, Fernando replied:

> I think it's mostly from living with her [Katherine]. Before I moved to New Orleans, I was living in Los Angeles, and there, you had the Latino bars, and you had the *gavacho* [a term used to refer to U.S.-born descendants of European immigrants] bars. I didn't speak a lot of English, but I would go to both. And when I went to the *gavacho* bars, I would try to speak with people, but it was hard.

Here, people are nice to me. Like today, I went to watch the game at Finn McCool's, and people felt sorry for me [Fernando had worn his Mexico soccer jersey at the bar]. And they put their arm around me and bought me drinks. (unstructured interview 2014)

In New Orleans, Fernando feels it is easier to negotiate the cultural divide that separates him from his host society, and he attributes this ease, in part, to what he perceives as a more welcoming attitude among residents. He is not alone in this feeling. Carolina also ascribes to New Orleans a familiarity with Latinos that she feels is rooted in the city's cultural ties to the Caribbean:

There is something that is unique to New Orleans that is friendly to Latinos—the culture, the architecture, the way that there is this very Latin and Caribbean nature of the city that seems familiar to them. (structured interview 2014)

I find Carolina's words interesting because they appear to contrast with her earlier statement that New Orleans was not a Latino-centered city. While seemingly contradictory, her impression that New Orleans was not Latino-centered stems more from the relatively low percentage of Latin American immigrants in the city before Katrina, while her impression of the city's friendliness may be attributed to the Caribbean, French, and African cultural influences that shaped new Orleans as a creolized space, which she finds affectively familiar.

Returning to Fernando's case, it is worth noting that he had already made strides to cut across those perceived differences he felt separated him from non-Latinos, such as visiting *gavacho* bars in Los Angeles, before moving to New Orleans. His experiences in New Orleans also offered additional opportunities for mimetic exchanges with non-Latinos with whom he made friends and eventually established intimate relationships. Furthermore, his perceived ease in establishing intercultural relationships in New Orleans may be partly credited to his developing English language skills. Following up on his comment about the differences between bars in Los Angeles and New Orleans, I asked him whether he frequented bars like El Rinconcito that catered to new Latino arrivals. He replied:

I've gone there, but you'll sit down, and then a woman will come up to talk to you, but they turn out to be *ficheras* [female entertainers who charge a fee for dancing and providing company for men at bars]. And, you know, it's not a good idea to be doing that. And I don't want to be paying a woman just to talk to her, so I don't go to those bars that much.

Living with her [Katherine], I've also changed. You know, now I do things I wouldn't do in Mexico like wash dishes and help with the laundry. Back home, if my friends saw me doing that they would say, "Mandilón, te pega la vieja!" [Coward, your old lady hits you!] (unstructured interview 2014)

Life in New Orleans, and the relationships in it, is changing Fernando in ways that may not have occurred had he not migrated to the city. In Los Angeles, he felt alienated from *gavacho* culture, and in Michoacán, the policing of gender roles by other men would not allow him to shift his ideas about masculinity and the practices he considers indexical of it. At the same time, Fernando's body carries a memory and a collection of affinities he has neither shed nor forgotten, but they are the foundation upon which he is building a new form of personhood in Louisiana, one that will become the medium upon which he will creolize as a person throughout his life.

It is also worth noting that Fernando's transformative process does not amount to a relinquishing of his nationalized identity or his identity as a Latino. His wearing of the Mexican national soccer team's jersey to a New Orleans Irish pub is particularly telling in this regard. For scholars of transnational migration (Basch, Schiller, and Blanc 1994), however, these processes of creolization shift a person's embodiment in such a way that they experience difficulty reintegrating into their communities of origin when they return, for the communities themselves are not static or unchanging and their members undergo their own transformative processes in the absence of those who left. Finally, creolization is not unidirectional. Katherine will also change through her life experiences with Fernando and yet will maintain her identity as a New Orleanian.

While mimesis is sometimes used to refer to an act of direct imitation, it has also been used to refer to mutually responsive and co-constitutive affective interactions between people. Anna Gibbs (2010), for example, has used the instance of a parent playing peekaboo with a child to show how interpersonal interactions that require a person to accommodate to specific affective expectations (e.g., when it is culturally adequate to smile) shape people's experience of affect and their embodiment. Relationships like Katherine and Fernando's, I would argue, can be thought of in the same way, because they require a careful dance of accommodation on the part of both partners that results in new quotidian practices for them—reconfiguring gender roles and identities and bringing about new embodied cultural expressions.

Creolization at Large

Over the course of the last century, U.S. foreign policy, military interventions, and economic interests played a key role in shaping the political culture and social life of many Latin American nations, particularly Central America and Mexico. U.S. policymakers and military strategists worked in collaboration with Latin American political elites in establishing production and trade relations that made labor, natural resources, and commodities accessible to U.S. markets. In many instances the creation of these transnational networks necessitated state violence to suppress critics, dispossess subsistence farmers, accumulate resources, and discipline populations into docile labor. The construction of this network, however, also had a notable impact in the formation of North American urban spaces.

Just as U.S. capitalist interests (in the form of fruit companies) traveled to Central America in search of commodities, Central Americans engaged these economic networks and traveled to southeastern Louisiana, becoming a part of the urban fabric of New Orleans. Central American immigrants became yet another group in the city's long history of immigration, joining the list of people who arrived from Germany, Ireland, Scotland, Italy, eastern Europe, and Russia in the nineteenth century. All of these groups were, at one point or another, perceived and affectively experienced as outside others who were not entitled to occupy the social and material spaces of francophone New Orleans, but they all creolized in ways that not only shaped the city by building ethnic enclaves but also created new expressions of their identities and forms of embodiment, neither assimilating nor remaining unchanged. Over the course of such creolizations, both established and new New Orleanians experienced shifts in the ways they affectively experienced the built environment and defined what it meant to be a city native.

Local politicians who commented that Latino workers either threatened or were averse to creole "culture" in post-Katrina New Orleans missed the point of creolization. Creolization, and people categorized as Creoles and criollos in colonial contexts, points toward embodiment not as a static, unchanging fact; instead, it represents the process of always becoming—a process that is as fluid, heterogeneous, and complexly interwoven as are people's social relationships in a cosmopolitan context. The concept of creolization also does a better job than the concept of assimilation in helping us think about the processes of culture change that inevitably occur in a world that is increasingly interconnected. Through creolization places such as New Orleans, the northern coast of Honduras, and Michoacán become entangled via the flows of capital, people, and material culture that form the twenty-first-century world.

8. To Love a Small Town

THE POLITICAL ECOLOGY OF AFFECT
IN THE MIDDLE MISSISSIPPI

For the engineers who work on the Mississippi River's water-channeling system, the Missouri city of Cape Girardeau marks an important spot. South of the city lies an extensive and robust network of federally funded levees that are rated as being able to withstand hydro-meteorological conditions—precipitation, snow melt, river water levels—that are so rare and extreme that they have a one in five hundred chance of mani-festing on any given year. North of the city, the river's levee system is a patchwork of levees built by local boards and the U.S. Army Corps of Engineers, with some being locally funded, constructed, and maintained, and some being federally supported. The latter levees are built to vary-ing specifications and with different auxiliary flood protection mecha-nisms such as pumps and drainage systems. Given these distinctions in river engineering and flood protection systems, engineers prefer to make the distinction between the Upper Mississippi (the river and its flood-plain north of Cape Girardeau) and the Lower Mississippi (the river and its floodplain south of Cape Girardeau). Still other experts like to add a third demarcation, the Middle Mississippi, which is said to be the stretch of river between St. Louis, Missouri, and Cairo, Illinois.

Behind the patchwork of levee systems in the Upper and Middle Mis-sissippi, towns, communities, farms, and households are situated within the river's floodplain and are susceptible to periodic flooding. Some of this flooding can be catastrophic, claiming households, livelihoods, and, on occasion, human lives. While floodplain residents, civil engineers, geologists, and other scientists who work on floodplain management agree that the flooding of these households and communities is unde-sirable, at times they disagree about how to address the situation.

Some flood mitigation specialists insist that hydrologic systems and rivers have agency. They periodically overflow their naturally formed levees and change their path, and people and societies must work with these systems and allow them to do what they have a propensity to do. Scientists and experts who espouse this perspective often rally around the saying "Make room for the river." In some instances, such an approach to floodplain management translates into a diminished investment in the construction of structural flood protection mechanisms like levees and a distribution of settlement patterns and land use practices that are in tune with the ways rivers act. The implication of this approach for people and communities in flood-prone areas is that their current living and home construction practices are no longer feasible. In some of these instances, mitigation experts recommend that people relocate their homes and communities to locations outside floodplains so that with more limited human involvement their current places of habitation can revert to floodplain ecologies.

While advocates of this approach to floodplain management see it as rational, floodplain residents in the Upper and Middle Mississippi region often resist it, and many refuse to abandon their towns and communities in search of higher ground. To outside observers and some mitigation specialists, this refusal seems, at times, illogical and difficult to understand. As one flood mitigation specialist told me in an ethnographic interview, such floodplain residents "are in absolute denial" (fieldnotes 2013). But the ethnographic work I have conducted among flood-affected communities in the Midwest since 2011 shows that what may seem a simple and logical solution to flood risk is a much more complex matter, and if mitigation experts do not directly address this complexity in their community relocation recommendations, then these plans run the risk of being seen as unfeasible and even socially disruptive by floodplain residents.

But what is this complexity I speak of? The ethnographic examples I present in this chapter demonstrate how understanding the reasons people live in flood-prone areas, the ways they respond to relocation recommendations, and the feasibility of actually relocating people and communities demands a recognition of the intimate relationships between national and regional development policies and practices, the ways people make space, the varying ways people relate to one another (i.e., sociality), and the bodily experience of emotions and attachments. Collectively, I refer to these relationships as the political ecology of affect.

Relocation, in its most basic sense, involves a movement of people from one space to another, but space, as I have shown in previous chapters, is by no means a neutral backdrop of human action. Rather, space emerges out of people's interactions with the material world that are further nuanced by culturally contingent ideas, intentions, and conventions. Space, as it is configured through people's relationship to other people and things, is the context where human beings experience the body and emotions in particular ways and therefore is also a place that is part of who they are. Although relocation may seem simply a move in space, it is a process that suspends those relationships that shape people as persons and can therefore destabilize their identities and disrupt the built and social environments that evoke their sentiments of familiarity, comfort, and home.

Space manifests not solely through the actions of people who live in floodplains but also through broader development policies and practices that are conceived in distant institutional localities and can play an important role in determining which spaces are viable and which are not in contemporary North America. Telling a community to resettle is simple. Relocating a community in a way its residents find meaningful, comforting, and sustainable in terms of their livelihoods is not. Understanding the political ecology of affect is key to devising flood mitigation policies that help floodplain residents and experts make informed decisions about resettlement, as well as relocation projects that make sense to communities that undergo this process.

To explore the political ecology of affect in the Middle and Upper Mississippi, this chapter asks several questions: Why do people live in places that are deemed hazardous by flood mitigation specialists in the first place? What development processes lie in the background of people's settlement of the Mississippi floodplain? What kinds of attachments to other humans, spaces, and environments do people devise over the course of their lives in the floodplain? And when floodplain residents do decide to resettle, what spaces are feasible for the relocation of their communities in the broader context of early twenty-first-century North America?

The Ethnography of Relocation in the
Middle and Upper Mississippi

This chapter is based on ethnographic research I have conducted in a number of Illinois towns and communities located in the Mississippi River floodplain since 2011. My involvement in this project began after snowmelt from the Upper Midwest and spring precipitation levels reaching 300 percent of customary rainfall caused flooding and required the U.S. Army Corps of Engineers to dynamite levees along the Ohio and Mississippi Rivers in May of that year. The purposeful destruction of these levees was a standard operating procedure meant to relieve pressure on the overall river system, thus flooding sparsely populated areas, and to protect major cities and settlements.

Among the towns that flooded was Olive Branch, Illinois (fig. 17), a community of approximately nine hundred people located just south of Cape Girardeau and northwest of the Illinois city of Cairo, where the Mississippi and Ohio Rivers meet. Following the catastrophe, members of a Natural Hazards Research and Mitigation Group based in the Department of Geology at Southern Illinois University–Carbondale (SIUC) approached the people of Olive Branch and proposed the idea of relocating the community to a higher elevation. Before the 2011 flood, Olive Branchers opposed relocation, but the damages sustained in this last disaster drove many to consider it.

According to mitigation group workers, immediately after the catastrophe, 90 percent of the residents polled during town hall meetings agreed to partake in a voluntary resettlement process. Olive Branchers, however, had one specific requirement: the relocation had to be a collective process. Existing FEMA buyout practices engage flood-affected households on an individual basis and with a focus on homeowners. These southern Illinoisans, in contrast, felt a sense of familiarity with one another based on their longtime residence in the area and on their shared identity as small-town residents. This sentiment is similar to the *confianza* of flood-displaced Cholutecans discussed in chapters 2 and 3.

Olive Branchers feared FEMA's approach toward relocation would create a situation where people would purchase or rent new homes in various nearby towns and cities, thus breaking the spatial proximity

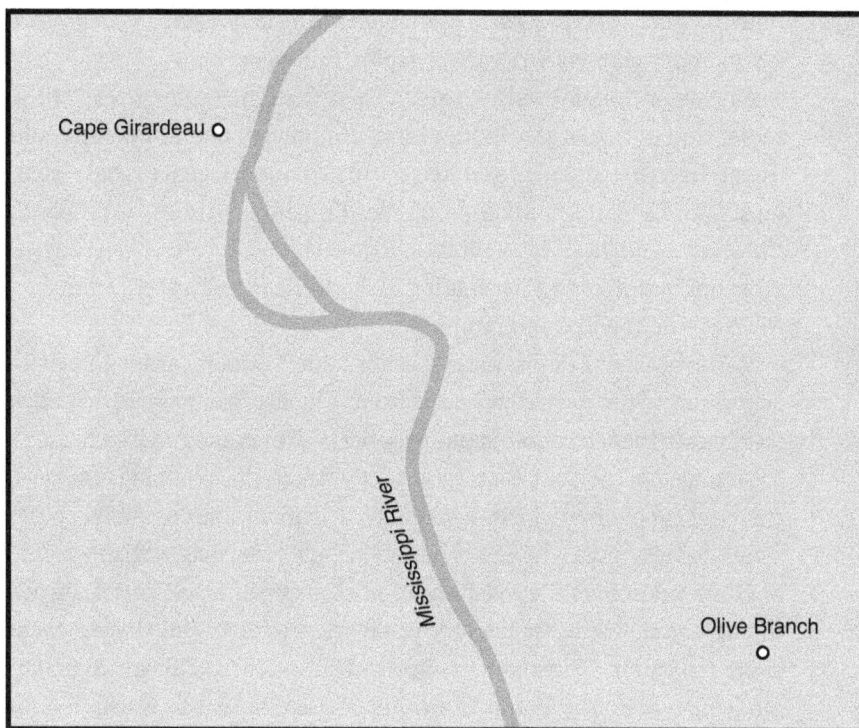

FIG. 17. Greater Olive Branch area. Courtesy of author.

they felt was key to maintaining the daily practices and social relations that created a sense of community. In their request to relocate collectively, Olive Branch residents made it clear that the social relationships among town residents were what they cared for most in the disaster recovery process. This situation emphasized once again that stranger sociality, as it is assumed in FEMA buyout programs, is at odds with the ways people experience community and imagine relocation.

One of the challenges Olive Branchers confronted was that their town was an unincorporated settlement and therefore lacked the official political representation—that is, a mayor and a city council—required in FEMA buyout procedures. To address this issue, representatives of the SIUC Mitigation Group offered to have the university play the role of local government in applying for a buyout, which would only partially fund the relocation process. Thus began a long and tedious journey that is still under way as I write this chapter, four years later.

As part of the university's role in the resettlement, members of the mitigation group organized an interdisciplinary team of faculty and researchers that included architects, urban planners, geographers, geologists, and social scientists from Western Illinois University, Lehigh University, and siuc. My invitation to join this team came from the two other anthropologists participating in this initiative—Heather McIlvane-Newsad and David Casagrande.

My involvement in this project has included working as an anthropological consultant for architects and urban planners assisting Olive Branch residents in the relocation-planning process. This ethnographic endeavor has entailed participant observation activities during planning meetings, charettes (public participation–planning activities), and community organizational meetings. The work has also included traveling with the people of Olive Branch as they visited other disaster-affected communities throughout the Midwest in search of experiences that would inform them about relocation. Christopher Higgerson and Katherine Ray, two undergraduate research assistants from siuc, assisted me in this research. This applied anthropology project also involved the completion of over sixty formal ethnographic interviews with residents of Olive Branch and comparable towns in Illinois, including Jacob, Grand Tower, Valmeyer, and Beardstown. Heather McIlvane-Newsad, David Casagrande, and their research assistants conducted the majority of the interviews, and authorship of this text belongs to them as much as it does to me.

Now I Understand Why You Live Here!

When I began working on this project, I mentioned my plans to visit Olive Branch to a fellow social scientist. My friend and colleague scoffed when I spoke of the small community: "They're nothing but a bunch of rednecks who now want 'Big Government' to bail them out" (unstructured interview 2012). While discussing the project with one of the architects involved in the relocation, I sensed a similar disdain on the part of academics for the residents of this southern Illinois town. "The houses are not much to look at," I was told. "It's just pole barns and ranches with vinyl siding for the most part" (unstructured interview 2012).

Olive Branch is not a cosmopolitan center. It is a small community founded in the late nineteenth century that was hit hard by the global economic crisis of the 1970s. Many families had to sell or foreclose their farms, and this period began a process of steady population decline for many North American rural communities (Adams 2002). Today Olive Branch is a town located along the sides of Illinois Highway 3, and non-residents drive through without giving it much thought. For some Olive Branchers, in contrast, the community evokes very different affective reactions. More than a year after the flood, the SIUC Mitigation Group organized a planning charette meant to generate ideas about how and where to resettle. Architects, urban planners, and social scientists from private practices and universities from across the Midwest, as well as community members attended the event. One of the residents in atten-dance, Jeffrey Lingle, a man in his forties, stood up, introduced himself, and said, "I am here to see what we can do to get Olive Branch back to what it was" (fieldnotes 2012). Jeffrey's voice cracked as he spoke the last words. Tears streamed from his eyes, his body convulsed with sobs, and he bowed his head as he sat down, not wanting to make eye contact with the rest of the crowd.

As I got to know Jeffrey better, I came to understand his display of emotion for a place that evoked sentiments of either elitist resentment or disdain among some of my colleagues. Before the 2011 floods, Jeffrey and his two older brothers were renowned in the town for their prowess as goose hunters. Larry, Jeffrey's older brother, had multiple titles as the state goose-call champion. The head of the Lingle family was Jeffrey and Larry's mother, Melinda, who was widowed more than a decade before the flood. She also owned a multiunit structure adjacent to Horse-shoe Lake, which, in the mid-1980s, was known as "The Goose Hunting Capital of North America."

In times gone by, nearly 150,000 geese descended on the lake, attract-ing hunters throughout the United States. The Lingles adapted the mul-tiunit structure to serve as housing for out-of-town hunters and made a significant part of their livelihood from hosting and guiding visiting goose-hunting enthusiasts. By providing a unique experience for their guests, the Lingles carved themselves a corner of the goose-hunting market. During hunting season the Lingles invited the visitors to their private homes, which were adjacent to the guests' quarters, and they spent evenings sharing goose call secrets and local hunting lore. This shared intimate experience introduced outsiders to a world of social relationships unique to the town. As I came to know Jeffrey and his family better, I developed an appreciation for how his sense of self was intimately tied to his labor as a hunting guide and how his identity and affective experience of Olive Branch were products of co-constitutive relationships between environment, people, and development history.

In recent years a number of changes in the regional ecology—some suspected of being related to climate change and others to changes in development and farming practices—have kept the geese from arriving in such large numbers. Today only 35,000 or so geese are thought to pass through the county during hunting season. This change has significantly affected the livelihoods of Olive Branchers who rely on hunting tourism. Nevertheless, Jeffrey and his brothers have continued to earn a living by hosting and guiding hunters who still come to Olive Branch looking for geese.

Jeffrey's personhood, I began to realize, could not be divorced from his lived experiences in the human-environment relations that make Olive Branch what it is. The historical political ecology of Olive Branch played an important role in shaping his body, making him the kind of person who weeps as he remembers this small town in its heyday of small family farms and hundreds of thousands of migrating geese. Resettlement, for Jeffrey, is not a simple move in Euclidian space; it is a task that poses a potential rupture of those relationships between the people and the environment that shaped his sense of self. Jeffrey's public display of emotion reminded me, once more, of the powerful relationships between affect, sociality, and the built and natural environment I had observed over the course of my work as an anthropologist of disaster. This chapter is about tracing those relationships and connecting them to broader political economic processes that sometimes create their possibilities of being and also foreclose them.

Not all Olive Branch residents are hunters, however, and their reasons for resisting the idea of relocation vary. Not so far from the Lingles' hunting camp at Horseshoe Lake is the house of Jackie Dunn's family. Born and raised in nearby Thebes (another small southern Illinois town), she planned to spend her elderly years in a similar rural community. When the Dunns purchased their house, the area had not flooded in fifty years and was spared by the Great Flood of 1993. When their neighbors discouraged them from obtaining flood insurance, and their insurance agent indicated the house was not affected by the Great Flood of 1993, they interpreted these statements as an indication that they were not at risk of flooding. "If it did not flood in '93, it won't flood," she remembers being told.

The Dunns bought that specific house so Jackie's husband could fish on a regular basis. The Dunns' house sits less than fifty feet from Horseshoe Lake's shoreline, and their living room features a floor-to-ceiling window that gives an impressive view of a waterscape with large cypress trees that have moss hanging over their limbs. Following the flood, Jackie and her husband refused to move. When members of the SIUC Natural Hazards Research and Mitigation Group visited their home and inquired about their decision to stay, Jackie waited until they were in her house and looking out the window to state her reason for not leaving. Gesturing toward the view of the lake, she asked, "Would you want to leave? Would you really?" (fieldnotes 2012). This was not the first time I had entered Jackie's house. On a previous occasion, I had visited with an architect from Chicago who was collaborating with the SIUC Mitigation Group in planning the community's resettlement. The architect had expressed reservations about the Dunns' decision to stay, but when he entered the house and saw the same sight, he exclaimed, "Now I understand why you live here!" (fieldnotes 2012).

The Dunns and the Lingles are just two of the hundreds of families who live in Olive Branch, and although both families differ in their reasons for either having misgivings about or entirely objecting to relocation, they also illustrate the ways people and both natural and built environments become intimately related in such places. For Jeffrey Lingle, Horseshoe Lake is inextricably tied up with a hunting culture that the lake makes possible; this hunting culture, at the same time, makes the lake what it *is* for Jeffrey. This man's sense of being, his gendered identity, and his affective experience of the body are constituted through his relationship to Olive Branch's surrounding environment that involves both social relations and the material world tightly wound together.

For the Dunns, life in Olive Branch is also a continuum of human-environment relations and social relationships among people whose personhoods are shaped by daily life in the small agricultural villages of southern Illinois. Certainly Jackie foregrounds the lake's impressive beauty as a subjectivity-transcending reason for living where she does, and visiting architects and researchers often agree that the lake is a sight to behold. Nevertheless, my ethnographic work among Olive Branch residents suggests that the reasons many of them lived in the area until 2011 were far more complex than the allure of its natural beauty. The Dunns, as with the Lingles, also partake in social and kin relationships that collapse the division between nature and society; they tie the material environment to social life, making Olive Branch into the place it is. Relocation can mean not only severing relationships to this material environment but also rearranging relationships among the people.

The Dunns and the Lingles are not the first families to live in the Middle Mississippi floodplain, and looking at the history of development practices in the region can help shed light on how spaces such as Olive Branch and people's attachment to them become possible in the first place. In the remainder of this chapter, I look at how the area surrounding the Middle and Upper Mississippi became permanently habitable and what possibilities exist for Olive Branchers and others when they decide to relocate after catastrophes. Both concerns show how an evolving capitalism plays an important role in shaping but does not fully determine the kinds of places and communities that are possible in North America.

The Historical Political Ecology of Attachment in the Middle Mississippi

The material landscape of the Middle and Lower Mississippi today is very different from what it was 130 years ago. Up to the Civil War, levee construction along the river's banks was, for the most part, a piecemeal affair, with local levee boards and landowners taking responsibility for constructing the structures to protect settlements and agricultural lands. The more significant events of the antebellum period were the creation of the U.S. Army Corps of Engineers in 1802 and the acquisition of new territory under the Louisiana Purchase in 1803. Population growth in the agriculturally productive area of the river delta soon followed (USACE 2014).

The expanding settlement of the delta welcomed the introduction of the steamboat in the Lower Mississippi, and commercial activity grew. By mid-century, 187 steamboats operated in the area (Camillo 2012; USACE 2014). While early French colonial settlements in the Lower Mississippi stressed levee construction to protect the settlements and plantations, the arrival of the steamboat and the commerce it made possible increased the importance of navigation—that is, ensuring the river had a continuous channel to allow commodity-carrying vessels to move. The country's transition from a mercantilist colonial economy to an incipient capitalist one featured quantitative and qualitative shifts in the ways people and governments engaged the materiality of the river and created spaces where settlers lived, had life experiences, and developed unique attachments to social and material environments.

The Civil War took a toll on this patchwork levee system as both sides of the conflict diverted resources usually devoted to river engineering to the war effort (USACE 2014). This continued neglect resulted in a catastrophic flood in 1874, leading Congress to establish the Mississippi River Commission—a governing body meant to bring together U.S. Army and civilian engineers—in 1879. The commission's task was "to transform the Mississippi River into a reliable commercial artery, while protecting adjacent towns and fertile agricultural lands from destructive floods" (Camillo 2012, 1). From its inception until 1927, the commission followed a "levees-only" approach toward channel construction and flood protection, and this practice not only created more spaces for the settlement of farming communities in the river's floodplain but also enhanced flood risks.

During the late nineteenth century, a key characteristic of the Middle Mississippi's levee system was a gap located on the west bank south of Cape Girardeau that allowed the river to expand into the St. Francis River basin in Missouri in times of potential flooding (Camillo 2012). The levees-only approach to river engineering culminated with the leveeing of this area in 1909. The river engineering efforts of the late nineteenth and early twentieth centuries transformed the floodplain, which had previously been subject to seasonal inundation, into highly fertile land that attracted new residents. Olive Branch was founded in this period, and its original footprint lay just outside the current hundred-year flood line, whose probability of flooding is 1 percent in any given year, according to the U.S. Army Corps of Engineers' calculations.

In 1927 the levee-focused approach toward river management reached a critical point. Heavy precipitation and snowmelt put excessive pressure on the system of closed-off spaces where the river could overflow, leading to more than two hundred levee failures, the flooding of 16.8 million acres of land, and 246 confirmed deaths. The flood's devastation prompted Congress to approve the 1928 Flood Control Act and charged the Mississippi River Commission to implement the Mississippi River and Tributaries (MR&T) project, which took nearly eighty-five years to complete. Rather than continue the levees-only approach that had partly shaped the 1927 disaster, the MR&T project applied multiple river-channeling and flood-mitigating mechanisms. In 2012 the Mississippi River Commission's historian, Charles Camillo, described these mechanisms:

> The engineering features include an extensive levee system to prevent disastrous overflows on developed alluvial lands; floodways to safely divert excess flows past critical reaches; backwater areas to store surplus floodwaters and reduce pressure on the levee system; channel improvements to increase the flood-carrying capacity of the river; channel stabilization features to protect the integrity of the levee system and to ensure proper alignment and depth of the navigation channel; and tributary basin improvements, to include levees, headwater reservoirs, and pumping stations, to maximize the benefits realized along the main channel by expanding flood protection coverage and improving drainage into adjacent areas within the alluvial valley. (Camillo 2012, 5)

The history of engineering practices in the Mississippi River, from the Gulf of Mexico to Cape Girardeau, Missouri, demonstrates that the river is anything but a natural object. The Mississippi is a place where human practice and desires (for transportation channels that generate commerce and for new agricultural lands) have met materiality. They have produced a tangible environment that is neither fully of people's making—water and soils have agency and sometimes behave in ways that lead to outcomes people deem catastrophic—nor fully "natural," given that the people's presence has changed the hydrology of the region (Pickering 2008).

The engineering of the Lower and Middle Mississippi has both reflected the unfolding of capitalist development in the United States over the last three hundred years and played a role in shaping what that capitalism looks like. While the river's own emergence as a combination of human practice and material agency could not be fully predicted by its engineers (the undesired and unexpected outcomes of human-environment relationships, the floods of 1874 and 1927 were by no means planned events), it also created new possibilities for relationships among people and between people and things. Over the course of the nineteenth century, cities and towns such as Olive Branch were founded and thrived in the floodplain spaces that became possible through the construction of the MR&T, and their residents developed attachments to natural, built, and social landscapes.

After the 2011 floods, Camillo celebrated the last eighty-five years of river engineering and the accomplishments of the U.S. Army Corps of Engineers by crediting its projects with preventing what could have been a more damaging flood, one surpassing the 1927 event. He wrote: "Since its initiation, the MR&T program has brought an unprecedented degree of flood protection to the approximate 4 million people living in the 35,000-square-mile project area within the lower Mississippi Valley. The nation has contributed roughly $14 billion toward the planning, construction, operation, and maintenance of the project. It has proven to be a wise investment that has prevented more than $478 billion in flood damages—a $34 return for every dollar invested" (Camillo 2012, 5).

Camillo's history of the MR&T project, however, leaves out the spaces that these feats of river engineering created and the lives that took place in them during the last century. Just as the MR&T project was coming into maturity during the second half of the twentieth century, some Olive Branch residents began to construct dwellings in the hundred-year floodplain, counting on the perception of flood security that the U.S. Army Corps of Engineers' projects created. As noted previously, when Jackie Dunn purchased her house at the shore of Horseshoe Lake, her future neighbors and insurance agent informed her the area was not heavily affected by flooding and that flood insurance was not necessary.

We were told this never flooded. It was fifty-something years old, and it never had water in it. So we bought it, and seven months later we had water in it after we remodeled it. We asked about the insurance. State Farm said, "You don't need flood insurance. It didn't flood in 1993 [the last major flood in the Upper and Middle Mississippi before 2011], and it's never gonna flood again." So we took the owner's word. (structured interview transcription 2013)

For social scientists interested in the relationship between environment, culture, policy, and practice, the last 215 years in the Middle and Lower Mississippi demonstrate how space is not merely a neutral backdrop where social action takes place but also what comes into existence the moment people (with their culturally situated intentions and desires) engage the world's materiality (Biersack 1999; Ingold 2000; Lefebvre 1992; Low 2011; Mitchell 2002; Pickering 2008). Perhaps more important, a space comes into being through the interplay of practices not only by the people who live directly in it but also by the actions of those who are removed from sites such as Olive Branch in space and time. This insight is particularly important, as discussions about flooding and resettlement are often laden with moral condemnation for people who live in disaster-devastated communities. Over the course of my ethnographic experiences in New Orleans, Central America, and the Midwest, I have had many conversations with students, colleagues, and friends who routinely ask why people chose to live in a disaster risk zone and why their tax money should be spent on bailing out those who made such a seemingly irresponsible decision.

Discussions about flood mitigation and resettlement often reduce the reasons why people live in a particular place and resist relocation to individualistic ahistorical terms and ignore the broader social and space production processes at work in the making of "at-risk" communities. If space and communities come into being in relation to policy and development processes that extend beyond their boundaries and that are not necessarily within the complete control of their residents, then how are we to reflect on the sociopolitical conditions that make resettlement communities viable? To put it another way, if the original settlers shared responsibility with broader processes of technological innovation, emer-

gent capitalist markets, and post–Civil War nation-building policies—
what often goes by the name of "progress"—for creating the conditions
that allowed Olive Branch to exist as a community, then what kinds of
processes are at work in making the spaces of relocated communities
possible in the early twenty-first-century Midwest?

In the sections that follow, I explore the latter query by reviewing the
experiences of Olive Branch residents during the partial relocation of
their town. This examination shows that in discussing resettlement (and
community resilience, for that matter), some disaster mitigation experts
often ignore the more extensive sociopolitical forces at play in putting
(and keeping) communities on the map. Instead, conversations usually
take a tone that places responsibility for the success of relocation efforts
on the disaster survivors themselves while overlooking the broader
agents at work in making specific places feasible. This review, I hope,
calls attention to the more extensive relationships between policy, policymakers, and disaster-affected communities involved in making recovery possible for Olive Branchers and others.

The (Im)Possibility of Relocation

The idea of community relocation can conjure images of transplantation,
where an organism is moved and adapted in its entirety from one context
to another while remaining by and large unchanged through the process.
The movement of human communities, however, never involves such a
simplistic change in location. Instead, it usually produces dramatic transformations in the relationships between people and their surroundings,
turning the process of relocation into an ecological change fraught with
a high potential for discontinuities and with arrangements that the displaced consider undesirable (Cernea 1996; Guggenheim and Cernea
1993; Macías 2009; Oliver-Smith 2009; Scudder and Colson 1992).

After the 2011 flood, Olive Branch residents began a collaboration
with faculty and researchers from siuc with the intention of imagining
possibilities for their town's resettlement and recovery. As part of this
joint endeavor, the siuc Natural Hazards Research and Mitigation Group
organized a tour of midwestern towns of comparable size that had experienced disasters and were in varying stages of recovery. The objective
of this multiday activity was for a select group of civically engaged Olive

Branchers to generate ideas about relocating their town and to think about the directions the reconstruction of the community could take. Over the course of a week, university faculty shared a bus with six town residents who formed the core of a nascent group of grassroots organizers concerned with the town's fate after the floods.

Our tour included visits to Valmeyer, Illinois, which was catastrophically flooded during the Great Flood of 1993 and resettled to a higher elevation; Greensburg, Kansas, which was devastated by a tornado rated EF5 on the Enhanced Fujita scale in 2007 and reconstructed in the same site; and Joplin, Missouri, which was also struck by an EF5 tornado in 2011. During these visits, Olive Branch residents spoke with community leaders and members of these localities about what was reconstructed, how it was done, what resources were involved, and how they were secured. Of the three disaster-affected locations, Valmeyer and Greensburg were comparable to Olive Branch in scale and organization. At the time of their respective catastrophes, these towns had populations of 1,500 people or fewer, were located in rural areas, and were experiencing a population contraction due to shifts in the U.S. agricultural economy during the second half of the twentieth century. The following section provides vignettes of the visits to Valmeyer and Greensburg, detailing the kinds of relationships between community leaders and state and federal agencies that facilitated their recovery, the kinds of spaces these relationships made possible, and the feasibility of similar recovery trajectories taking place in Olive Branch.

VALMEYER, ILLINOIS

In 1993 Valmeyer was a small community of approximately nine hundred people located on the east bank of the Mississippi River floodplain and a little more than a hundred miles northwest of Olive Branch. Although the town was located in an area used for agricultural production, it was also only twenty miles from the southernmost extension of the city of St. Louis. Counting 325 homes within its legal boundaries, the settlement was first incorporated in 1909 and had experienced a number of flood events in the 1940s until a levee construction project completed in 1950 granted the town more robust flood protection. The levee was part of a regional system that included St. Louis, and it had a section (what river engineers call a fuse plug levee) that was designed to fail and relieve water pressure upstream in case catastrophic flooding conditions threatened the larger city.

Because the 1993 flood damaged the majority of housing and business structures in Valmeyer (some estimates claim 95 percent), the town mayor led a push to relocate the community. The chosen relocation site was a five-hundred-acre farmland property located at an elevation four hundred feet higher than the original settlement. But relocating Valmeyer was not a guaranteed process. Federal Emergency Management Agency funds for relocating communities, or "FEMA buyouts," are handled through programs that only reimburse property owners for a portion of the market value of their homes (sometimes as low as 75 percent), and that amount is often far below the cost of new construction. Additionally, buyout programs do not provide funds for the infrastructural development of relocation sites, such as building roads and installing basic services like power and potable water.

In Valmeyer's case, the FEMA buyout paid a total of $23 million to homeowners, while the relocation site cost $22 million. The town's successful relocation required a monumental effort on the part of community leaders, who took responsibility for raising the additional funds necessary for developing the new site. Because of the land's cost, Valmeyer residents found themselves forced to obtain new mortgage loans, and those who either could not or would not bear such a financial burden (primarily the elderly) had to find homes in other urbanized areas. While today Valmeyer's population is approximately a third larger than it was in 1993, the town also lost a third of its original members during the relocation. Roughly half of Valmeyer's current residents did not live in the town at the time of the flood, and only six hundred of the nine hundred or so of the original neighbors live in the area (Leonard 2013).

Securing funds for Valmeyer's relocation required community leaders to embody a number of specific social and organizational skills that are not necessarily common among some residents of towns like Olive Branch. Dennis Knobloch, the mayor at the time of relocation, was a successful insurance salesman and familiar with bureaucratic processes. His business networking skills made contacting and negotiating with state and federal officials seem natural to him. In 2013 in his office as county clerk for Monroe County, Illinois, former mayor Knobloch recalled the complex balancing acts he had to carry out to negotiate the town's relocation:

> Imagine having a conference call with fifty people representing twenty-five different federal and state agencies, each with their own requirements for funding, and their requirements sometimes contradict each other. That's the kind of thing you have to do. (structured interview 2013)

One advantage of the new relocation site was that it was located upon a bluff whose base had been carved out through mining operations in the early twentieth century. The remaining mines were excavated in such a way that they could be easily converted into temperature-controlled storage facilities. The town government obtained the rights over the mine and established leases with the U.S. National Archives and major food distribution companies, securing a source of income that would help offset the cost of making the site habitable. Another advantage of the new site was its proximity to St. Louis. Over the course of the relocation, town officials marketed the new Valmeyer as a bedroom community for St. Louis residents, hence the significant proportion of new neighbors who inhabit the town today. While the original Valmeyer had become a feasible living space in a broader late-nineteenth-century and early twentieth-century agrarian economy, the new Valmeyer exists as an effect of spatial practices of contemporary North America, in which middle-class residents have shaped suburbs as a means of escaping the perceived ills of inner-city life such as crime and living among ethnic minorities (see Low 2003).

Over the last two decades, Valmeyer has earned a reputation as what Mary Delach Leonard (2013) of the *St. Louis Beacon* called "a model for other floodplain communities." Indeed, members of the SIUC Mitigation Group pointed to Valmeyer as the quintessential example of successful community resettlement and urged other at-risk communities to replicate its results. Anthropological perspectives on embodied class formation, however, suggest that forms of personhood such as that of Mr. Knobloch (for whom the bureaucratic and networking skills necessary for successful relocation seemed instinctual) are not a given but are ways of being that require careful lifelong cultivation. Moreover, they are key mechanisms through which social groups (and classes) produce themselves and their social bodies (Bourdieu 1977; Foucault 1978), or what Pierre Bourdieu (1977) has called history turned into nature.

While the work Mr. Knobloch did for his community to realize the relocation is commendable, how replicable the case of Valmeyer might be is questionable. Olive Branch residents quickly noticed this and came to terms with it during our visit. Because Olive Branch was an unincorporated community, it lacked a local government and a clear set of community leaders to represent the town and make its case before funding agencies. Of those community members who emerged as local leaders during the recovery, none saw themselves as comparable to Mr. Knobloch in terms of professional bureaucratic skills or embodied "cultural capital," although they possessed important knowledge about their town and were successful within its own economy of family farms and recreational hunting. The case of Valmeyer highlights how FEMA and other federal agencies do not recognize communities as objects of intervention or assistance and how they prefer to deal with individual homeowners as recipients of aid (buyouts). What is more, as other anthropologists of disaster have observed (Adams 2013; Browne 2013), the bureaucratic processes that communities must undergo to obtain federal assistance carry a built-in class bias: they inadvertently favor people who have bureaucratic literacy skills, which members of predominantly working-class communities do not necessarily possess.

GREENSBURG, KANSAS

In 2007 Greensburg, Kansas, was a town of approximately fourteen hundred people. Like Olive Branch, the town's past was tied to an agricultural economy that shifted in the second half of the twentieth century from small, labor-intensive family farms to larger, more mechanized agricultural operations. Consequently, the town was experiencing population contraction and an uncertain future when, on May 4 of that same year, a powerful EF5 tornado 1.7 miles wide leveled the town, killing eleven people.

During its reconstruction, community leaders and assisting urban planners successfully branded Greensburg as a prime example of "green reconstruction." The town's recovery featured the installation of wind turbines that produce more energy than the town consumes, with the surplus energy being exported to other communities. The town also benefitted from the construction of energy-efficient public buildings (schools, city hall, hospital) whose innovative design catches the eye of visitors and architects alike.

While Greensburg is not necessarily a classic example of community relocation, it may very well demonstrate what Deborah Davis Jackson (2013) has termed *community dysplacement*. Although the community is located in the same geographic location, the 2007 tornado destroyed all but one familiar landmark—the town's courthouse—and its prolonged recovery cost it about a third of its population, predominantly elderly residents. The town may not have moved physically, but it is not necessarily the same place it was before the catastrophe.

The SIUC Mitigation Group selected Greensburg as a place to visit because of its innovative approach to recovery. Over two days, Olive Branch community organizers and their accompanying entourage of anthropologists, architects, and journalists met with city government officials and residents to discuss the organizational challenges of planning and executing a reconstruction process. We learned from Mayor Robert Dixson that planning for the town's green recovery added further delays to the reconstruction time line, as it involved more extensive design activities as well as a more complicated fund-raising process to finance the building of costlier energy-efficient structures.

For Greensburg's elderly, the prolonged recovery did not seem feasible. Many opted not to reconstruct in the old site and looked for other housing alternatives in nearby towns. Disaster reconstruction and resettlement in both Valmeyer and Greensburg, then, resulted in population loss and exclusion that would not have occurred without their respective catastrophes and unique approaches to recovery. If space involves not only a geographic area but also the social relations that take place within it—thus giving it meaning, memory, and value—then we can say that Valmeyer and Greensburg became significantly different localities over the course of recovery. This development stands in contrast to the vision of resettlement that the Olive Branch residents originally entertained. They defined their town not merely as a geographic space but as a collection of social relations, and they imagined relocation as the transplantation of these connections among people.

Conversations with prominent Greensburg residents also shed light on the kinds of relationships between federal government agencies, politicians, and disaster survivors that are conducive to the reinvention of a community and the feasibility of green recovery. The Greensburg catastrophe occurred almost two years after the federal government's mismanagement of its disaster response to Hurricane Katrina. As the Greensburg hospital administrator explained to Olive Branch residents, the town secured the funds for constructing a new state-of-the-art medical facility because the Bush administration felt pressured to demonstrate its ability to help communities recover from disasters and not because of established federal disaster response protocols. "The only reason we were able to build this hospital is because the Bush administration told USDA [U.S. Department of Agriculture] to make it happen," the administrator commented (fieldnotes 2012). It was also telling that Greensburg's high school was reconstructed as an energy-efficient facility with brick and wood recycled from New Orleans households, even as the latter's devastated communities such as the Lower Ninth Ward remained without such public services.

As noted in chapter 2, the important political leverage of the presidential administration on federal agencies to facilitate the reconstruction of a community demonstrates that spaces become viable as a result of political relationships that extend well beyond the geographical boundaries of any given community. Olive Branch residents, in contrast, did not have such leverage after the 2011 floods, and simply imagining an energy-efficient community was not going to be sufficient to secure the resources they needed to relocate in a way they found meaningful. Southern Illinois is notably de-prioritized in state politics, with the greater Chicago area enjoying the lion's share of state funding allocations. Not only does Olive Branch have low visibility in terms of state governance, but also its catastrophe was overshadowed by other more media-friendly disasters that took place in 2011, like the tornado in Joplin, Missouri. Following the Illinois flood, for example, in a move that could have helped garner public support for the small town's relocation, a popular cable network's home-remodeling television show was planning to film one of its episodes in Olive Branch; however, the show's producers found the Joplin disaster more sensationally viewer friendly and canceled the Olive Branch appearance.

Olive Branch Four Years after the Flood

As I write this chapter, the ordeal of community relocation is far from over for many Olive Branchers. In the flood's immediate aftermath, town residents were hesitant to participate in a FEMA buyout program for a number of reasons. For one, people were worried that the program's treatment of town residents as individual homeowners would lead to a scattering of their community should people find individual housing arrangements in nearby towns and cities. Equally worrisome for Olive Branchers was the management of FEMA buyout funds. Because Alexander County government is renowned in Southern Illinois for its corruption and financial mismanagement (Muir 2015), the SIUC Mitigation Group obtained the flooded families' support for the buyout application only after the researchers and faculty assured them that the university would step in for county government and act as the direct administrator of FEMA funds. In 2012, however, higher-level university administrators became concerned about the possible legal fallout of providing fiscal oversight. Previous partnerships between the university and Alexander County government had gone awry, resulting in audits by the Internal Revenue Service. To avoid future liability, the university withdrew from its role as fiscal overseer, disheartening those Olive Branchers who had already committed to the buyout.

Difficulties with the buyout also emerged when securing a land parcel for relocating the flooded households. Initially a nearby property that belonged to a prominent farmer was identified as a suitable site. During initial negotiations, the farmer quoted one price, but upon hearing about FEMA's possible involvement in the relocation, he later quadrupled the property's value. The change in price would have left Olive Branch residents in a similar situation as their Valmeyer counterparts, forcing them to obtain mortgages for new construction. The change in price proved to be a deal breaker, and the flood victims no longer considered the site an option. In April 2015, only twenty flood-damaged homes had been officially closed through the FEMA program, and more than a hundred properties remained waiting on their final buyout transaction, a process that could take up to a year to complete (Kurwicki 2015).

Conclusion

Community relocation is often recommended as a means of mitigating flood risk for those communities located in the floodplain of the Middle and Upper Mississippi. The case of Olive Branch, however, reveals that relocation is by no means a simple process and that the state and federal agencies that are supposed to assist communities through it often fall short of providing the necessary help to execute what disaster survivors deem successful relocation projects. Even in the few exemplary cases such as that of Valmeyer, the fate of relocation projects seems to fall on the shoulders of community leaders, creating a bias for those communities whose representatives have the cultural capital often associated with upper-middle-class households. In this way, relocation programs, as they are currently managed by FEMA, seem to reiterate class distinctions rather than earnestly provide the aid that disaster-affected communities need to recover. This is not an isolated phenomenon. Anthropologists who work with flood-affected communities have recognized similar tendencies in the way U.S. government agencies manage aid programs (Browne 2013; Maldonado et al. 2013).

The case of Olive Branch also requires us to reflect on the nature of space in floodplain communities and the spatial-affective challenges of relocation. As I have noted before, the space of this town became possible only within a broader process of regional and national development that engaged the Mississippi River's materiality and was closely linked to a type of commerce that was contingent on specific technologies (steamboats, levees) and took shape as North America transitioned between a colonial mercantilist economy and an incipient capitalism. But this was only part of the story. The leveeing of the Mississippi River to protect farmland and create a continuous channel to efficiently move commodities also created spaces where human lives took place, and people shaped forms of personhood and affective dispositions in relation to this environment, which was as much material and natural as it was social.

Discussions of relocation often downplay the affective attachments people have to the places where they live and assume that space is merely a neutral backdrop to human action, a preexisting grid where a population may move from point A to point B. The ethnographic study of community relocation shows a very different picture. The spaces of disaster-affected communities do not merely exist; indeed, they become possible in webs of co-constitutive relationships between policy, the environment's materiality, historically and socially situated desires, and human practice—or what some anthropologists would call a political ecology (Biersack 1999). In disaster relocation programs, taking seriously the attachments people have to the human and human-material relationships that manifest in their space is to recognize a very tangible reality of people's experience of life, environment, and recovery.

9. Rebuilding It Better

THE ETHICAL CHALLENGES OF DISASTER RECOVERY

On a crisp and bright day in May 2012, community organizers from Olive Branch met with various city officials and residents of another disaster-devastated town, Greensburg, Kansas. Accompanied by academics, researchers, and journalists, the Olive Branchers had traveled eleven hours by chartered bus from southern Illinois to see how members of a comparable community had taken on their own reconstruction process. As described in chapter 8, on May 4, 2007, Greensburg was almost completely leveled by an EF5-rated tornado. Prior to the catastrophe, Greensburg was experiencing population losses similar to that of Olive Branch. Once a predominantly farming community with some industry, the small Kansas town's population had steadily declined in the preceding decades as younger generations sought employment and life opportunities in major urban areas. In 1990 Greensburg counted 1,782 residents. By 2000 this number had decreased to 1,560, and in 2007, on the eve of the storm, Greensburg had a population of 1,265 (U.S. Census 2016).

As in the case of New Orleans and many other communities devastated by disasters, a number of community leaders, government officials, and assisting academics saw the catastrophe as an opportune moment to reflect on the town's long-term development trends. Concerned with Greensburg's population contraction, this group of local leaders and external advisers considered approaches to reconstruction that would shift the town's pattern of resident loss to one of resident gain. Among the approaches they examined, the idea of rebuilding the town as a prime example of energy efficiency, renewable energy production, and sustainability stood out. Through the collaboration of planning teams, private energy companies, and substantial financial support from the U.S. Department of Agriculture, Greensburg followed a path toward green reconstruction.

In May 2012 Greensburg had wind turbines that produced more energy than the town needed for its own consumption and a collection of buildings that received LEED (Leadership in Energy and Environmental Design) platinum ratings, which are used to recognize buildings built to the highest energy conservation and efficiency standards available at the time of construction. The structures include a city hall, a library, a county museum, a hospital, and a high school. In a short video about the town's reconstruction produced by the popular newspaper *USA Today*, Mayor Robert Dixson spoke about what he considered to be the meaning, importance, and reasons for "rebuilding green":

> Part of being green and sustainable and being resilient is also the fact that you leave it better than you found it. So we had an opportunity to do what our ancestors did when they founded Greensburg in 1886 and built this community—the chance to leave a legacy for future generations as our ancestors left for us—and that's what truly being green and sustainable really is . . . None of it would have happened if it wasn't for the resiliency and the determination of the citizens of this community. To say, "This is our community, this is our town, and we are rebuilding," now that is true leadership of the people that stayed, built new homes, got their businesses back going—that is what we take a lot of pride in. Yes, we have a lot of LEED buildings, a lot of energy-efficient buildings; we have a wide variety of architectural designs; we are a living laboratory, but those don't mean anything if it wasn't for people who stayed and inhabit these buildings. (Quinn 2013, 1)

Following our visit of the various energy-efficient structures and our conversations with town officials, administrators, and a select group of residents, our group of Olive Branchers, academics, and researchers moved to the newly built town hall for our final activity, to hear Mayor Dixson give an account of the recovery process. He told us that immediately after the catastrophe, the town's political leaders, assisting academics, and urban planners from nearby state universities convened to consider the town's possibilities for recovery. The idea of rebuilding green emerged out of these meetings, but this approach to reconstruction required a relatively longer planning and implementation process.

Mayor Dixson acknowledged that the people of Greensburg did not unanimously accept the decision to rebuild green. Advising Olive Branch residents on how to navigate the kinds of tensions and differences that invariably manifest in disaster recovery, the mayor said, "No matter what decision you take, you are always going to have naysayers, but you have to stick to your plan" (fieldnotes 2012).

The naysayers in his case were Greensburg residents who felt an urgent need to regain a sense of normalcy after the catastrophe, to get on with their lives, to collect their insurance settlements, and to begin the reconstruction of their homes as soon as possible. Like Olive Branch, Greensburg had a greater proportion of elderly residents in comparison to many other U.S. urban areas. Most of these elderly residents lived on a fixed income, had already paid off their properties, and did not consider taking on a new mortgage as a viable alternative. They also believed they were living the last years of their lives, and spending such years in a protracted reconstruction process did not seem desirable. Sticking to the plan of green reconstruction may not have solved Greensburg's long-term sustainability problems. In 2013 the U.S. Census Bureau recorded a population of 785 residents for the town, meaning that the disaster and its aftermath greatly exacerbated the community's problem of population loss.

Mayor Dixson's statements raised a number of questions concerning ethics and governance in disaster reconstruction. By using the term *ethics*, I mean to call attention to the ways people involved in disaster recovery conceive of what is a right and proper way of governing, making policy decisions, and putting policy into action. Following my experiences in Olive Branch and Greensburg, I continue to reflect on these questions: Who is a rightful member of a community? For whom should reconstruction be tailored? Are elderly residents and other highly vulnerable sectors of a population not entitled to consideration over the course of reconstruction? Who defines what it means to rebuild "better," and what is at stake in such a definition?

In this chapter, I detail how these questions led me to consider how Michel Foucault's and Giorgio Agamben's observations about the premises and unresolved tensions of modern forms of governance provide a helpful point of departure (but not a ready-made explanation) for thinking through the ethical complications of disaster recovery. In the following pages, I review their commentaries on the transformation in the reasons for and legitimizations of governance since the eighteenth century. Then I put their observations to an ethnographic test by exploring their relevance for a case study from Chiapas, Mexico, involving the Sustainable Rural Cities Program. In the process, I continue to document the ways disaster survivors invoke affective experiences in their evaluation of disaster reconstruction programs. Finally this chapter sets up a number of questions concerning the leap from critique to practice. In chapter 10 I not only address them but also provide five specific recommendations to help readers navigate the ethical dilemmas of rebuilding better.

Anthropological Reflections on the "Better"

As I have shown in the various case studies featured in this book, whether it is the resettlement of flood-devastated neighborhoods in southern Honduras, the recovery planning in post-Katrina New Orleans, or the FEMA buyout of a flooded town in southern Illinois, certain social actors routinely see post-disaster contexts as opportune moments to "leave it better than you found it." But defining what "better" is, these case studies also show, is never a matter of simple pragmatism where the better is self-evident to all who are involved. The better, as Foucault (1970) and Paul Rabinow (2005) would have us recognize, is always defined within a historically and sociopolitically contingent circumstance where cultural, epistemic, economic, and political factors cannot be neatly separated from one another. Consequently, the better is never as obvious or commonsensical as Mayor Dixson suggests. Its definition and enactment could be said to be a form of power at work, where some arrangements among people and things are credited with self-evident superiority over others although, in practice, this superiority is often questioned by disaster-affected populations and difficult to evince ethnographically.

In post-Mitch Honduras, municipality officials, community leaders allied with local political elites, and architects from Nacional de Ingenieros imagined the better as a master plan laden with modernist assumptions about the relationship between regimented space, minimal aid, and people's embodied subjectivities. The plan's inherent assumptions—that Cholutecan disaster survivors were alienated subjects who related to one another via stranger sociality and would flourish when given minimal aid packages—were not shared by the flood-displaced residents. The rigid application of these assumptions, furthermore, inhibited the ability of disaster survivors to experience a sense of recovery. The better, in this instance, was also imagined in the form of "integrated solutions" within a "functional [capitalist] housing market" by program managers from the U.S. Agency for International Development, but these integrated solutions and the housing market were not conducive to the creation of residential structures that displaced Cholutecans could recognize as homes or of socio-spatial relationships that evoked a feeling of *hallarse*.

In New Orleans a number of reconstruction actors and institutions—for example, Mayor C. Ray Nagin, the Bush administration, local developers, gentrifying resident constituencies, the Housing Authority of New Orleans, the Department of Housing and Urban Development, and expert planners—saw the extensive flooding triggered by Hurricane Katrina as a key moment for improving the city. The "better" city imagined in documents such the Unified New Orleans Planning proposal, however, was ultimately a mechanism for the investment, circulation, and replication of capital by strangers who lived among strangers and not for a space of familiar sociality among New Orleanians who were shaped as sensing and emoting beings over the course of life histories in the city's socially produced spaces. This way of imagining New Orleans also featured a shift in the relationship between the public and private sectors in which the former increasingly funneled contracts, funding, and the responsibility for public services to the latter (Adams 2013). The better of the UNOP plan prioritized the life of capital and neoliberal subjects who moved through the city as ahistorical people relating to each other via stranger sociality over the social lives of many subaltern New Orleanians. The recovery practices prescribed in the plan resulted in the prolonged displacement of residents who lived in heavily devastated neighborhoods such as the Lower Ninth Ward and more than 110,000 residents who self-identified as African American (Mack and Ortiz 2013).

Social scientists who research disasters often remark on the post-catastrophe moment as an opportune occasion for recognizing the relationships between people, practice, policy, and the agency of the material environment that engenders disasters (Bankoff and Hilhorst 2001; Hoffman and Oliver-Smith 1999). From this perspective, the post-catastrophe moment also carries the possibility of addressing and transforming such relationships over the course of reconstruction, effectively mitigating disaster. Nevertheless, the aforementioned case studies indicate that recognizing the political ecological relationships that give disasters their form and magnitude is not a given in post-catastrophe contexts. In the case studies I have presented, the ways urban planners, local government officials, and NGO program managers defined the better did not so much reflect a self-evidently rational improvement upon a given situation; instead, in a historically, politically, and culturally contingent way—and epistemic space—they imagined what better is. What I mean by epistemic space is the imagination (and the socio-material conditions that made such an imagination possible) that allows a social actor to conceive a particular reconstruction policy or practice. The case studies I presented also demonstrate how these epistemic spaces had strong neoliberal and modernist underpinnings that engendered policies and practices based on inherent assumptions about the natures of people and social well-being that were not shared or desired by the populations they were meant to assist. These policies and practices also failed to deliver on their promises of modernization and recovery through the privatization of social services.

Discourses of rebuilding it better have a tendency to articulate the visions and desires of experts and political and socioeconomic elites while marginalizing the voices of those people who actually experience the brunt of a catastrophe's effects: the elderly residents of midwestern U.S. rural towns, *clase obrera* Cholutecans, and sociopolitically marginalized New Orleanians. Perhaps most important, visions of rebuilding it better also implicitly state which kinds of lives are worth living and which are not after a catastrophe. In New Orleans, the life worth living articulated in the UNOP plan was the life of strangers who lived among one another as strangers in a landscape of newly constructed, colorful, and renovated architecture. The life worth living was also the life of capital, and the plan worked to enhance its reproduction, growth, and well-being. Implicitly, the UNOP plan also articulated what kind of life was undesirable. The revitalization of public housing that was agreed upon by UNOP planners and HUD and HANO administrators without any public input contributed to the prolonged displacement of many New Orleanians of modest means. Moving them not only disrupted the social lives they led but also curtailed the contribution they made to the production of the city's urban landscape.

In Honduras, meanwhile, the distribution of aid in Limón de la Cerca— for example, the layout of the site's master plan, the housing design, and the municipality's policies on land parcel ownership and random distribution—ignored the social relations and spatial arrangements that characterized the neighborhoods of disaster-displaced Cholutecans before Mitch and the important role of these relations and associated arrangements, such as child care and household safety, in their everyday lives. The construction of Limón de la Cerca made the implicit statement that the way of life in pre-Mitch *barrios* in Choluteca was one not worth living, while life in the reconstruction site, with its modernist design and stranger sociality, was worth it even as this locality became a site of violence, alienation, and death.

Sovereign Power and the State of Emergency

A more in-depth exploration of the "lives that are and are not worth living" complication of disaster recovery stands to inform broader discussions in the social sciences about the unresolved contradictions of modern forms of governance and sovereign power. As discussed in chapter 1, Foucault once observed that Niccolò Machiavelli and other sixteenth-century political philosophers entertained ideas about the legitimacy and obligations of sovereign power that are difficult to justify in today's liberal democratic societies. Foucault claimed their ideas about the rights, ethics, and responsibilities of sovereign power were based on the notion that the sovereign had two bodies—the first being the monarch's physical body and the second being a metaphorical mapping of this body on their estate. The primary responsibility of the sovereign, this line of thought held, was to foster the life and well-being of this same estate. Consequently, treatises on governance such as Machiavelli's *The Prince* counseled the sovereign on how to expand and maintain his or her second body.

In Foucault's analysis, the estate constituted the lands, the people, and the infrastructure of the territory controlled by a sovereign, but the sovereign's obligation toward his or her subjects did not include providing those things needed to maintain their living bodies. Other institutions, such as the church, were charged with the obligations of giving alms to the poor and establishing hospitals to care for the ill. Because the people who lived within the sovereign's estate were considered an extension of the sovereign's body, however, any assault on those subjects constituted an assault upon the sovereign, who reserved the right to put to death the transgressor.

As I also discussed in chapter 1, Foucault noted a shift in the ways sovereign power, governance, and the relationship between monarchs and their subjects were conceptualized and legitimized between the seventeenth and eighteenth centuries. A key element of this shift was that the focal concern of governance became the care for human populations as biologically living entities. Accompanying this trend, Foucault saw a proliferation of disciplines that would eventually become public health, urban planning, and the social sciences, which focused on the subject of "man." Together these disciplines provided the knowledge necessary for creating and maintaining a milieu where the life of people as biologically living and economically active beings could be fostered (Foucault 2004). Foucault christened this modality of sovereign power "biopolitics" and noted its shift from the power over death (as in the sovereign's right to punish those who violated their estate) to the power over life.

Biopolitics, Foucault believed, comprised the epistemic ground upon which the policies and practices of many twentieth-century state institutions were conceived. What was particularly troubling about biopolitical power, however, was how the modern state's injunction to foster life led to justifying the use of deadly force to protect the lives of those who were deemed to be members of a "national race." Biopolitics, then, defended the killing of those people who threatened what were considered normative forms of life. In the foucauldian project, for example, Nazi Germany represents the pinnacle of biopolitical state modernity, and the genocide of Jewish, disabled, and other so-called undesirable populations became a prime example of the violence biopolitics makes possible (Agamben 1998).

More recently, Agamben (1998) has taken on Foucault's project, further exploring the philosophical ground upon which modern forms of biopolitical sovereignty and governance can be imagined. At the heart of biopolitics, Agamben maintains, is a way of thinking about people as *homo sacer* (sacred man) and bare life. *Homo sacer* is a person who is ritually removed from society and can be killed but not sacrificed; thus, this human's body is allegedly exorcised of meaning and is simply reduced to a fleshy mass, or bare life. In the modern Western imagination, Agamben argues, the killing of *homo sacer* by a sovereign power does not present the breaking of a ritual or a societal or moral taboo.

The origins of *homo sacer* lie in classical Roman civilization, where soldiers facing the possibility of death in battle—a death without the necessary rite of passage for a proper afterlife—subjected themselves to a ritual that transformed them into a life that could be killed but not sacrificed during the socially liminal period of warfare. *Homo sacer*, Agamben claims, provided the philosophical ground upon which the bare life of modernist thought and biopolitics could be conceived. Modern ways of thinking involve the desire to see the world independent of cultural values, to see a world of "things in themselves," and *homo sacer* lent an epistemic prototype for just such a way of imagining people.

For Agamben, the imagining of people as biopolitical bare life manifests most clearly when contemporary state governments declare states of emergency. Under the state of emergency, routine political decision-making processes that involve representative lawmaking bodies are suspended, and executive officials wield extraordinary powers. Among these extraordinary powers, the executive has the ability to kill those who are thought to present a threat to a biopolitical state's subjects without pursuing the customary juridical procedures of habeas corpus and a trial by jury. The U.S. military and Central Intelligence Agency's drone program in the Middle East, Africa, and South Asia provides a key example of this phenomenon. In this case, the state of emergency is extended to everyday life, and the executive branch of the U.S. federal government claims the right to extinguish the lives of people (both U.S. citizens and noncitizens alike) who are deemed a threat to the American public without the customary due process.

I would like to make the case that disasters also present a juridical-political situation in contemporary state societies where elected officials and agents of governmental institutions assume special powers under the justification that a state of emergency exists. Disasters present a complication, however: the decisions that these powers make do not so much concern the killing of bare life to protect a biopolitical social body (although rumors about such decisions often circulate in disaster contexts) as they do about what kinds of lives are deemed worth living and which are not. Returning to Foucault's observation that modern states are often qualified by the imperative to create milieus that attend to the economic and public health of biopolitical life, I now focus on exploring the ethical, social, and practical complications that arise when elected officials and governmental institutions approach post-disaster reconstruction as an opportunity to create such environments.

While Agamben's and Foucault's ruminations on biopolitical sovereignty help bring the unresolved ethical tensions of disaster recovery into relief, they also raise a number of questions that are best explored ethnographically. How do biopolitical ways of thinking about disaster recovery actually play out in practice? How do reconstruction actors in different settings interpret, enact, or reconfigure the biopolitical assumptions of modern governance? How do biopolitical ideas about sovereign power intermingle with already established governance practices such as colonialism and post-coloniality as they circulate at a global scale? To explore these questions, I introduce one last ethnographic example from Chiapas, Mexico.

I collected the ethnographic evidence presented in this chapter during a research sabbatical leave in the fall semester of 2013. Geographer Jesús Manuel Macías, anthropologists Fernando Briones and Joel Audefroy, and architect Nelly Cabrera—all of whom have conducted or directed research on the Sustainable Rural Cities Program in Mexico—assisted and advised me in this project.

Sustainable Development and the Resettlement of San Juan de Grijalva

In 2007 San Juan de Grijalva was a small rural community located along the banks of the Grijalva River in the northwestern municipality of Ostuacán, state of Chiapas, Mexico (fig. 18). Numbering a little more than

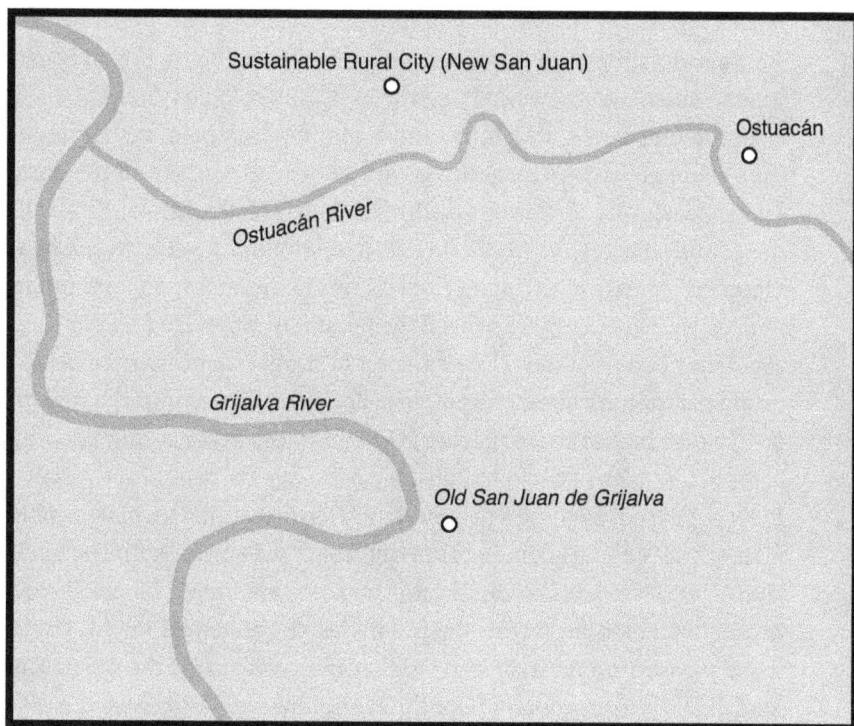

FIG. 18. Greater Ostuacán area. Courtesy of author.

four hundred, San Juaneros were predominantly small-scale farmers who practiced limited animal husbandry, sought wage labor on occasion, and supplemented their household diet by fishing and hunting. The community had eighty-seven households, and their homes were constructed in what architects often call the Chiapanec vernacular architecture: wood structures with large, covered front porches where people sit during hot daytime hours and with large living rooms where most daily activities take place. Housing structures, separated by agricultural fields and animal corrals, were dispersed throughout the landscape along the shores of the Grijalva River.

San Juan de Grijalva had only existed in its locality for two decades. Between 1979 and 1987, its section of the Grijalva River was heavily altered by the construction of the Peñitas Hydroelectric Dam, and the settlement was moved to a higher elevation to accommodate the river's new level. As the final stage in a broader system of four hydroelectric stations along the Grijalva River, the Peñitas Dam construction began in 1964 and extensively modified the area's hydrologic system.

In the final week of October 2007, an unprecedented 47.5 inches of rain fell in eight days and, along with poor management of the dam, put critical levels of stress on the hydroelectric facility, making the failure of the structure's retention walls a possibility. Just eighty-six kilometers northeast of the dam, the city of Villahermosa, a major urban area on the Gulf Coast, also faced the possibility of severe flooding, and the dam's potential failure threatened to trigger an even more devastating disaster. Although less violent flooding did occur in the region, at 8:32 p.m. on November 4, fears of Villahermosa's cataclysmic flooding came to a sudden end when a hillside on the eastern bank of the Grijalva River collapsed, creating a massive landslide measuring 1.9 billion cubic meters only sixteen kilometers upriver from the Peñitas Dam. The landslide's massive debris plowed into the river, creating a blockage 800 meters wide and 80 meters deep. This blockage helped relieve rising water pressure on the dam, and Villahermosa was spared the worst. The landslide's violent contact with the river, however, also created a 165-foot tidal wave that instantly submerged a significant portion of San Juan, killing twenty-five people.

The specific cause of the landslide remains debated to this day. According to official accounts, the massive land movement was caused by the interaction of geological and hydro-meteorological factors. Heavy rains soaked the area's hillsides, putting internal water pressure on geological layers, and the increased weight of loose topsoil resulted in a landslide. Former residents of San Juan de Grijalva, however, often counter this official explanation. They claim state agents purposely created the land movement to protect Villahermosa and simultaneously treated them as life that, to borrow Agamben's words, could be killed but not sacrificed. Edgar Córdoba (2012) from the University of Puebla conducted a survey of community residents and found, for example, 80 percent of San Juan's survivors believed the landslide was purposely caused. One of Córdoba's respondents explained the event: "The form the disaster took indicates it was intentional . . . the perfect and straight cut created by the mass of land makes us think it was provoked by an explosion. Something else that was suspicious is that, just following the disaster, there were many rescuers and helicopters. The truth is, they [state emergency organizations] behaved in such a way that we started asking 'why is the Government putting so much effort into this?'" (Córdoba 2012, 24, translation by author).

Other residents told Córdoba a more specific story. They said the landslide was triggered by a group of Israeli and North American ex-military surveyors who were exploring the area for multinational mining companies: "Before the disaster there was a great presence of foreigners who were supposedly here to work on reforestation. They were from Israel and they were also accompanied by Americans. We suspected their activities because they only went out at night, but we think that, in reality, they were interested in minerals because there is much uranium and petroleum in the area and we saw them several times speaking with Pemex [Mexican state-owned petroleum company] personnel. There were also various small planes flying around the area" (Córdoba 2012, 27; translation by author).

When I arrived in Chiapas in October 2013 to visit and speak with San Juaneros about the community resettlement process that followed the catastrophe, similar stories were still circulating. Gustavo Sanchez, a twenty-seven-year-old man who lived through the event, looked at me intensely during a conversation about his life in the resettlement community. He asked, "Are you not afraid to be walking around here asking about these things?" When I asked what he meant, as I considered my questions about everyday life in the resettlement site to be relatively innocuous, he replied that the San Juan de Grijalva landslide was a dangerous topic to broach because paramilitary operatives triggered the disaster. He continued,

> It was a group of gringos, and the captain was from Israel. I saw them a few times in Ostuacán [the municipal capital]. They had maps. It was a group of foreigners. And they said they were working on a reforestation project, but they were strange. They all had shaved heads; they bathed together, men and women in the streams; and they were always on horseback. (structured interview 2013)

Don Andrés Dias, a sixty-two-year-old former resident of San Juan de Grijalva, told a similar story:

Supposedly, the "disaster," which I put between quotation marks, was purposely created; it was not a disaster. There are neighbors who say it was created on purpose. On the days before, we saw small airplanes flying at low altitude. And the place where the land moved, the cut in the land is straight. A natural disaster disturbs the land and doesn't leave an even line. This landslide is straight. They put dynamite in it to save the state of Tabasco and also to extract minerals that were in there. They just didn't calculate well. That's why I put "disaster" between quotation marks. Twenty-six [the exact body count is contested] people died there, including my mother and younger brother. (structured interview 2013)

It is impossible to determine whether the landslide was purposely triggered. Two landslide specialists who have visited and assessed the site commented in ethnographic interviews that they could not, with complete certainty, discard this possibility. One explained:

What is interesting about this landslide is that it is much, much bigger than anything I have ever seen. Landslides that occur naturally are usually much smaller and more numerous. What is interesting about this one is that it is just one massive movement. (semi-structured interview 2013)

Other social scientists who have researched this disaster have offered alternative explanations that involve human actions although not intentional ones. Their explanations call attention to the environmental impact of the four hydroelectric projects constructed along the Grijalva River and of deforestation and to the inadequate management of the Peñitas hydroelectric station (CEPAL 2008; Córdoba 2012). A fellow anthropologist who has investigated this particular disaster, for example, suggested that the geological composition of the region, combined with dam construction and poor management, triggered the landslide:

The area has a limestone layer beneath the topsoil, and it is possible that poor management of water levels in the dam led to rapid changes in the water table, which helped loosen the limestone layer, causing the landslide. (unstructured interview 2013)

Narratives that explain the landslide as the result of intentional human action merit ethnographic attention, not because they are either verifiable or false, but because they reveal the kinds of anxieties that are unique to the sociopolitical circumstances of their narrators' living conditions (Briggs 2004). Approached from this perspective, such stories help orient the ethnographer's analysis when tracing the kind of global connections, governance practices, and hegemonic discourses at play in a disaster-devastated locality. The idea that San Juan de Grijalva could be destroyed in order to protect urban Villahermosa and to exploit mineral resources is a concern that is unique to people living in a situation where postcolonial, biopolitical, and neoliberal tendencies get inextricably entangled in the practices of state and federal governments, their agents, and their private sector associates.

What is postcolonial in this case is the legacy of Chiapas's status as a provincial region of Mexico; the state is characterized by an extractive economy as exemplified in the multiple Grijalva River hydroelectric projects. These programs do not provide a direct benefit for the "dispersed rural poor" (as Chiapas state agencies have represented the people of San Juan de Grijalva in the catastrophe's aftermath), but at the cost of extensive local environmental impact, they do generate the energy that powers Mexico's urbanization and industrialization elsewhere. Post-coloniality also manifests in the status of subsistence farmers, who are increasingly seen in national development policy as a vestige of the past that must be modernized and incorporated as wageworkers for the transnational labor market of neoliberal schemes such as the 1994 North American Free Trade Agreement and the 2001 Plan Puebla Panamá (a regional market integration and development project extending from Central Mexico to northern South America).

Finally, the idea that San Juan de Grijalva could be destroyed to protect Villhermosa articulates distinctly biopolitical anxieties in which a postcolonial modernist state sees subsistence farmers as lives that can be killed, but not sacrificed, for the greater good of protecting the region's major metropolitan area. I argue in the remainder of this chapter that although the specific causes of the 2007 landslide are debatable, the resettlement and reconstruction of San Juan de Grijalva have exemplified the entanglement of postcolonial, neoliberal, and biopolitical tendencies. Furthermore, in the name of "rebuilding it better," this reconstruction has dramatically transformed the ecologies of San Juanero households by altering the people's access to agricultural lands and the river's resources and their space-making practices. Thus, the reconstruction has brought about a disaster after the disaster. The narratives of San Juaneros that attribute the landslide to postcolonial, neoliberal, and biopolitical entanglements, then, may not be too far off in their analyses of the catastrophe at hand.

The Sustainable Rural Cities Program

The resettlement of San Juan de Grijalva was executed within a broader community relocation and regional development endeavor called the Sustainable Rural Cities Program that was managed by the Chiapas state government. Designed as a mechanism to combat rural poverty by relocating small communities to urbanized development centers, the program was based on the assumptions that rural poverty is the product of population dispersion and that the solution to social inequities requires a spatial transformation of Mexico's countryside. The program's organizational manual reads:

> The extreme poverty and the conditions of exclusion from a dignified life of thousands of Chiapas families *originate from population dispersion*. The public policy of the Sustainable Rural Cities has, as its end, the changing of this reality in Chiapas, bringing about a better territorial distribution of the population in relation to the potentialities of regional development in a framework of greater social and economic prosperity and of sustainability in the use of resources. The Sustainable Rural Cities are designed and built to meet these purposes, accomplishing the urban aggregation of the people who live dispersed in rural localities with high levels of marginality or in risk zones, in places where services of high quality are granted, where there is urban equipment and, above all, economic, social and human development. (Gobierno del Estado de Chiapas 2013, 1; translation by author, emphasis added)

The claim that spatial dispersion is the root of socioeconomic inequities in southern Mexico goes against established anthropological knowledge about Mesoamerica that identifies colonial processes of dispossession and ethnicity-based discrimination as the causes of poverty in places like Chiapas (Adams 1970; Carmack, Gasco, and Gossen 2006). More important, the recommendation that such inequities can be resolved by resettling dispersed populations is a reiteration of long-standing visions of territorial development that date to the sixteenth-century colonial period. As discussed in chapter 4, during the colonial era Iberian settlers viewed indigenous populations as a source of gratuitous labor, and the colonists established the governance practice of *encomienda*, which entitled European *encomenderos* to the labor and tribute of indigenous populations in exchange for their Christianization. Many indigenous communities lived in scattered villages and resisted the *encomienda* system by migrating away from European settlements, so the colonists developed yet another colonial project, the *reducciones*. These resettlement programs aimed at controlling indigenous populations, their labor power, and their tributes through forced resettlement in high population density areas (Carmack, Gasco, and Gossen 2006; MacLeod 1973).

What is novel about the Sustainable Rural Cities Program, however, is that it intertwines the colonial imagination that made the *reducciones* possible with the imperatives of a neoliberal development mind-set and the urgency of sustainability in the context of climate change. In June 2008 the governments of Mexico, Guatemala, Honduras, El Salvador, Costa Rica, Panama, Belize, Dominican Republic, Nicaragua, and Colombia agreed to collaborate in the Proyecto de Integración y Desarrollo de Mesoamérica (Project of Mesoamerican Integration and Development), which evolved out of its predecessor, the Plan Puebla Panamá (Mariscal 2009). Both plans called for a sustainable approach to natural resource management and regional planning that also would address population dispersion in southern Mexico (Audefroy and Cabrera 2014). Additionally, the Mesoamerica Project recommended that such sustainable development should be accompanied by the economic integration of the region's people, meaning they should participate in the transnational labor and commodity markets. What could be better?

Two years after the landslide, San Juaneros resettled to a Sustainable Rural City called Nuevo San Juan de Grijalva, which is located fifteen kilometers away from their original village and seven kilometers from the municipal capital of Ostuacán. The resettlement project was carried out via a public-private partnership in which multinational companies that specialize in consumer products financed recovery projects for housing and infrastructure in exchange for tax breaks. The resettlement zone was intended to do more than simply provide housing in a hazard-free environment; it was a social-engineering project designed to transform subsistence farmers and agricultural day laborers into wageworkers and entrepreneurs connected to a transnational system of commodity production and exchange. State authorities encouraged San Juaneros to sell their lands, and for the more humble farmers (those owning three hectares or less), those lands were now too distant to be solvent given the cost of transportation.

In the resettlement site families received minimal land parcels that did not provide the land necessary for subsistence farming, for the lots were occupied primarily by new housing structures (fig. 19). The strong economic core of the resettlement project, meanwhile, was given a veneer of environmental sustainability and energy efficiency. The community featured solar-powered public lighting and homes constructed with local sand and clay that had been compacted into construction blocks on site with special machinery. Sustainability, in this case, was ambitiously conceived by the Chiapas Institute of Population and Rural Cities as a quality that covered various domains of human experience.

As an alternative source of income for its residents, the Sustainable Rural City was outfitted with a dairy processing plant, greenhouses, chocolate manufacturing facilities, and a factory for producing uniforms and schoolhouse furniture. State authorities selected a small group of San Juaneros to manage these various enterprises, while the vast majority of relocated residents were hired as laborers. Initially the operation of these industries required the significant involvement of government officials. The uniform and furniture factory, for example, only remained in business because of the paternalistic relationship between the Chiapas state governor at the time, Juan Sabinas, and Nuevo San Juan. The governor funneled contracts directly to the factory, but the contracts no longer arrived once his term ended. In October 2013 the factory had been closed for nearly a year, and 150 former employees were still waiting for their final paychecks.

FIG. 19. New San Juan de Grijalva. Courtesy of author.

Other entrepreneurial projects also faced difficulties because San Juaneros did not have reliable connections with intermediaries who could link them to regional markets. Over the course of ethnographic conversations, San Juaneros detailed how greenhouse workers were defrauded of an entire harvest of habanero peppers when a middleman took the product and promised to return with their payment within three days but was never seen again. Despite this setback, the greenhouse was the only project that remained in operation in 2013. Reflecting on the failure of the resettlement site's various projects, resident Andrés Dias stated,

Going from one culture to another is not easy. It takes a lot of work. People who work farming in the countryside and have never been in business have difficulty going from one to another. I have never been behind a sales counter. I worked with my machete, but never in a store . . . And from one day to the other, you find yourself in a store, we are not used to that. I used to dedicate myself to agriculture. I planted maize and beans, and now where?

They [San Juaneros] have never been in this situation. They are farmers. The government must take that into account. Managing a company is not a part of my profile. They give people these greenhouses, and that might be agriculture, but that's different. We are used to working in the open air; that [the greenhouses] requires technical knowledge. There are no jobs here. This is what I call a White Elephant. (structured interview 2013)

Don Dias's words call attention to the importance of practice and embodiment in resettlement projects, especially those that propose the rapid and facile transformation of community economies, human ecologies, and people's dispositions. The practice of being a subsistence farmer, Don Dias argues, creates a kind of embodiment with affective dispositions (i.e., a predilection for working outdoors) and skill sets that are not easily replaced overnight or in a predictable manner. He was not alone in his concerns about the resettlement site. Carlos Martinez, a forty-one-year-old resident of the resettlement, made similar observations:

We were happier over there, where we were. Here, there is no work. You have to buy everything. We used to work in the factory, but they have not paid us, and it's been almost a year. It didn't work. It failed because of the government's mismanagement, and now I don't have my lands. It was fifty pesos a day to go, and my land was small, so I sold it. The majority have sold their lands. Those that have forty or fifty hectares, for them, it's still worth spending the money on transportation, but not for me.

Now you are beginning to see families where the man has to leave and go look for work somewhere else. I mean, look, today is Monday, and there's no work. The men have to leave. Now when we leave, we have to go all the way to Cancún or Merida to find work. If they resettled us, this was supposed to be a sustainable city, and it's a lie. You come here, and there's no work. (structured interview 2013)

The built environment of the Sustainable Rural City not only attempted to transform San Juaneros from subsistence farmers into wage laborers and business managers but also set out to change an entire human ecology that involved housing structures, affective sensibilities (what one finds pleasant), embodied ways of being, and people's relationship toward the material environment. Don Martinez continues:

> Over there, we planted maize, rice, bananas. We had bird corrals with turkeys and chickens. We had twenty or thirty large turkeys. Here, we don't have enough space. We can have seven or eight turkeys at most. Those who don't have to live here say this is the way to eradicate poverty. Here, we have people who go days without something to eat. Over there, we had work with the ranchers, and we planted our own maize and beans. Here, if I don't have a job, who gives me anything? Here, we have to pay for everything. Before, we could always trade a turkey for a pair of pants, and if we were hungry, we could go fish in the river. (structured interview 2013)

Rather than leading to sustainability, this "rural city" transformed self-sustaining farmers and agricultural day laborers into a largely underemployed, dependent, and migrant labor force. What is more, the transformation of the city's household ecologies went beyond the relationships between people, land, and animals that formed their livelihoods before the landslide; it also involved an attempt at reengineering the most intimate household spaces. A little more than four hundred housing structures were constructed in Nuevo Juan de Grijalva to house not only the displaced San Juaneros but also the members of more than three hundred families from eleven other nearby communities deemed to be at risk of hydro-meteorologically triggered disasters (landslides, floods).

The newly built homes, however, did not follow any of the established local patterns of home construction such as the Chiapanec vernacular architecture. Housing structures were not equipped with front porches, the main entrances were located to one side, and instead of having front doors, the main bedrooms were situated facing the street. These design features had a disruptive affective effect on San Juaneros, making it difficult for them to experience any sense of familiarity with the housing units. A few households were able to pool sufficient resources and alter the designs by constructing front porches and relocating the front entrances (fig. 20), but the financial hardships experienced in Nuevo San Juan put such remodeling projects outside most families' reach.

FIG. 20. Modified house in New San Juan de Grijalva. Courtesy of author.

Don Carlos Martinez was one of a dozen or so residents who invested in making modifications to his house:

> I've put about 10,000 pesos (US$900) into this house. We fixed it up—the front facade, the entrance. We made it bigger by adding a porch. Look where they put the kitchen and the bathroom. We also had to remodel the inside. We changed where the bathroom is, little by little. We didn't like the layout, but they left it like that. The front entrance all the way on the side of the house, and we didn't like it that way. And the houses were not waterproof. Water used to come in through the windows, and nobody likes it when it rains inside. But just like that, little by little. (structured interview 2013)

Housing design was not the only spatial transformation jarring the affective sensibilities of San Juaneros. Homes in the old San Juan de Grijalva were distanced from one another, separated by agricultural fields and house gardens, creating a sense of privacy for villagers. The overall design of Nuevo San Juan took none of these spatial practices into consideration. Planners laid out the homes so that rather than resembling a city, the community bore the aesthetic design of a suburb. While cities are often characterized by poly-temporal structures (buildings built at different moments in time whose architectural characteristics materialize a variety of spatial practices) with varied uses and purposes, the Sustainable Rural City imposed homogeneity by building identical private residences in close proximity to each other but distanced from structures meant for commerce or manufacturing.

As Don Dias explained, San Juaneros affectively experienced this design as places of discomfort and insecurity on a daily basis:

> It's not easy going from one culture to another. We are a rural people. Over there [in old San Juan], we were accustomed to getting home from work and laying on a hammock to rest. Now, I get home, and one neighbor turns on his radio and the other is arguing with his wife. There are also many fights now. You are resting, and there's the neighbor getting drunk, and that makes me feel a lack of security. (structured interview 2013)

Milieus, Affect, and Lives Worth Living

The Sustainable Rural Cities Program set out to dramatically transform the milieu of the people of San Juan de Grijalva in the hopes that such a change would connect them to regional and transnational networks of commodity production and exchange and would propel them from a status that the program's creators perceived as one of rural poverty to one of urban sustainability and economic prosperity. In rebuilding it better, the creators of the Sustainable Rural City's milieu attempted to reengineer almost every dimension of the lives of San Juaneros—from their relationships to the environment as subsistence farmers, to their relationships with each other as village residents, and to their affective experience of the built environment.

In the era of climate change and globalization, community resettlement is increasingly proposed as a doubly beneficial solution to problems of disaster risk and national underdevelopment (Marino 2015; Marino and Lazrus 2015). The United Nations Development Programme heralded the Sustainable Rural Cities as an exemplary case of how resettlement projects can combine environmental, neoliberal, and vulnerability mitigation agendas. This program proposed technological and entrepreneurial solutions for what were, at heart, historically and politically created inequities unique to colonial and postcolonial extractive economies within and across Mexico's national boundaries. Nevertheless, the resettlement project's sustainable development did not improve the lives of San Juaneros; instead, it disrupted every aspect of their community, from the public to the most intimate.

As with the reconstruction of Greensburg, Kansas, the resettlement of San Juan de Grijalva was based on planning decisions that made implicit assumptions about what kinds of lives were worth living and which were not. The institutional actors who imagined these two projects also made claims to have a unique grasp on what the better way of living was, a claim that imbued their policies with moral authority. The ethnography of disaster reconstruction, however, provides empirical documentation that challenges that moral authority and illuminates the unresolved ethical complications and practical challenges of governance in post-disaster recovery.

In several of the case studies reviewed in this book, reconstruction actors in positions of institutional or sociopolitical influence made policy decisions and engaged in reconstruction practices that significantly transformed the built environment and social relations of disaster-affected communities. The actors legitimized these practices and policies on the grounds that they would improve or enhance the resilience of cities, towns, and communities recovering from catastrophes. Their policies and practices upheld ideas about market rationality and modernization as solutions to disaster vulnerability and economic marginalization, but in case after case, they compounded the socio-material impacts of disaster and prolonged the catastrophes' consequences over the course of reconstruction and resettlement projects.

How do we go from critique to practice? What contributions does an affect-centered approach to disaster anthropology make toward addressing the ethical dilemmas and policy challenges outlined thus far? In chapter 10 I reflect on these questions on the basis of the presented case studies and provide five recommendations for practice that may help readers navigate the ethical and political complexities of disaster reconstruction.

10. The Anthropology of Affect and Disasters

FROM CRITIQUE TO PRACTICE

The anthropology of disaster lets us recognize that catastrophes take form as a result of human practices that enhance the socially disruptive and materially destructive capacities of geophysical phenomena and that inequitably distribute a catastrophe's impacts along the fault lines of a society's body politic. Disaster reconstruction policies and practices that propose market-driven recovery (the circulation of people and capital of the UNOP plan in New Orleans, the regional market integration of Nuevo San Juan de Grijalva, the housing market of post-Mitch Honduras) or the spatial modernization of disaster-affected communities (the regimented, homogenized, and minimalized spaces of Limón de la Cerca and San Juan de Grijalva) not only fail to address the root causes of disasters but also create realities that complicate the survivors' affective and socioeconomic experience of recovery.

In post-Mitch Honduras the modernist regimentation of space disrupted important social relations critical to local practices of household security and child care. In New Orleans the conceptualization of the built environment as a mechanism for capital investment and circulation both ignored and failed to address the processes of racialized difference making that historically engendered disaster vulnerability in the city. In San Juan de Grijalva the drastic transformation of subsistence agriculture and household ecologies to a market-integrated suburban "city" created conditions of economic insecurity and affective disorientation among displaced families. In all of these cases, disaster survivors invoked their affective experience—in terms of *hallarse*, comfort, appreciation, and feelings of security—as the criterion through which they assessed the relevance and efficacy of disaster reconstruction policies and practices.

While presenting the research in this book to colleagues at a disaster reconstruction workshop at Chongqing University, China, in 2014, a fellow anthropologist who has conducted research on the aftermath of the 2008 Wenchuan earthquake voiced a criticism. Disaster reconstruction, he noted, is a much more fluid process than my ethnographic descriptions suggested. It involves negotiations between program managers, local government officials, and disaster survivors, and the negotiations can blur the lines between these actor categories. Disaster reconstruction is also a process where expert discourses are actively appropriated and reconfigured by affected populations; therefore, they can also feature cultural production where new subject positions and identities become possible. I agreed that in all the case studies I had presented, the lines between actor categories did not remain fixed, and disaster-affected populations actively interpreted and sometimes appropriated expert discourses and imaginings of recovery.

The reconstruction process is certainly an instance of culture change and transformation. The case of Marcelino Champagnat in southern Honduras, for example, demonstrated multiple instances of negotiation between project managers and disaster-displaced Cholutecans that were conducive to creating a resettlement zone that the latter could recognize as mitigation and recovery. In New Orleans, neoliberal visions of the city's possible future captivated the imagination of gentrifying resident constituencies, who became key allies of expert planning teams in legitimizing so-called participatory recovery plans. In Limón de la Cerca and post-Katrina New Orleans, however, socio-material conditions that inhibited the subaltern disaster survivors' ability to affectively experience recovery arose when expert planners and NGO program managers mobilized expert knowledge and techniques of governmentality—for example, project assessment mechanisms of cost-benefit analysis, budgets, and capitalist rubrics of urban planning—as nonnegotiable matters in disaster recovery. For this reason I argue that while disaster recovery is certainly characterized by negotiation, discourse interpretation, and practice reconfiguration, it also has moments when expertise or claims to knowledge of the better are mobilized in ways that steer disaster recovery processes in directions that many disaster survivors find of little relevance to their embodied ways of being and modalities of sociality.

While people who experience disaster do not remain unchanged and their imaginations are certainly sometimes inspired by expert modernist and neoliberal visions of disaster recovery, they equally experience moments of social and affective disruption when they see their space and sociality-making practices undermined by reconstruction efforts. This book's intervention is important because affect is a bodily experience that is ever-becoming (as in the case of creolization) and yet mnemonic, for the feeling body is a repository of memory that is triggered by familiar presences, smells, tastes, and sounds. Furthermore, the experience of affect comes into being over the course of histories of practice within human and political ecologies. As the people of San Juan de Grijalva demonstrate, although they are capable of adapting to new conditions and willing to inhabit new spaces, they are also embodied sensing beings shaped through histories of practice, and these embodiments delimit both their possibilities and limitations as their communities are radically reconfigured in the era of climate change, increasing urbanization, free trade zones, and globalization. We must therefore be doubly guarded in our anthropological analyses of disaster reconstruction. We must not err on the side of representing a neatly divided world where people remain unchanged and imaginings of recovery are methodically mapped onto specific actor categories, but we must also not forget that impositions do take place and that disaster survivors often voice important critiques that are silenced through institutional knowledge-making practices such as budgets, recovery plans, and official reports.

But what specific recommendations for practice does an affect-centered anthropology of disasters have to offer? The critiques I have presented translate into five recommendations. While not specific prescriptions, these recommendations are orienting principles that may help practitioners reflect on the affective, ethical, and epistemic dimensions of disaster reconstruction.

1. Recognize the historical ecologies of disaster-affected communities and expert visions of disaster recovery.

First among these recommendations is the need to recognize the historical, material, environmental, sociocultural, and political contingency of both disaster-affected populations and the ideas and techniques of expert planners, government officials, and non-profit and NGO program managers. Both disaster survivors and expert visions of recovery are the products of human ecologies that, while connected to broader worlds, ideas, policy movements, and histories, also have unique particularities. In the cases I have presented, ideas about modernism, market-driven recovery, and integration into twenty-first-century globalized capitalism that institutional and governmental actors promote as ready-made techno-political solutions to the challenges of disaster recovery failed to create the socio-material conditions that helped those who are most affected by disasters reconstruct their lives in ways they found meaningful, comforting, and feasible.

2. **Ascertain the practices that enhance the socially disruptive and materially destructive capacities of geophysical phenomena, and devise disaster reconstruction practices that directly address them.**

The anthropology of disasters has a long history of demonstrating that a disaster's effects are often inequitably distributed along the fault lines of the body politic. Techno-political disaster reconstruction agendas that focus on logics of capital investment, modernization, or the rapid transformation of populations into neoliberal subjects that can be easily plugged into transnational circuits of labor, commodity production, and exchange routinely fail to address the fundamental human practices that give disasters their form and magnitude. In southern Honduras, modernist resettlement site design and strict adherence to cost-benefit analysis only exacerbated the socioeconomic vulnerability of disaster survivors. In New Orleans, the conceptualization of urban space as a site of capital investment and replication on the part of UNOP recovery planners failed to address the critical role of space- and race-making practices on the part of New Orleanians that historically limited the socio-spatial mobility of working-class African American residents. In both Greensburg and San Juan de Grijalva, approaches to sustainable reconstruction failed to address the broader policy and social forces that had led to population decline in midwestern U.S. communities and that had contributed to the socioeconomic marginalization of subsistence agricultural communities in many parts of Latin America, respectively.

To be relevant, disaster reconstruction policies must directly focus on the location-specific practices that inequitably distribute a catastrophe's social and material effects. In the case of Honduras, this effort means challenging the ways political power operates in postcolonial localities and the ways such political culture becomes entangled with development reconstruction programs. In the case of New Orleans, it means transforming the ways racial difference has a history of being made through quotidian spatial practices and not simply applying a logic of capital investment as a panacea for the marginalization of inner-city African American populations. In the case of Olive Branch and Greensburg, it means changing FEMA flood mitigation programs so that they engage communities, not just individual homeowners, and so that they provide disaster-affected individuals and families with the necessary resources to rebuild their homes and lives instead of a fraction of their property values that is insufficient for reconstruction.

3. Question the "nonnegotiables" of disaster reconstruction and practice epistemic flexibility.

In a number of the reviewed case studies, expert actors (planners, architects, NGO project managers) and government officials perpetuated the deleterious effects of disasters when they invoked specific principles and techniques of disaster recovery—for instance, cost-benefit analysis, capital investment logics of urban reconstruction, modern and neoliberal urbanism—as certainties that were not up for discussion. The attribution of a nonnegotiable status to these principles and techniques simultaneously denies their particular historicity and ecology and grants them a status as readily universally applicable matters of fact, or what some might call best practices. Despite these claims, ethnographic research demonstrates that these principles and techniques enabled the creation of social, spatial, and material conditions that disaster survivors could not affectively experience as recovery. In contrast, when reconstruction actors in institutionally influential positions (e.g., NGO program managers) were willing to negotiate knowledge-making and program-assessment instruments such as budgets (as the case of Marcelino Champagnat in southern Honduras demonstrates), those instances proved critical in executing disaster reconstruction projects that made sense and served those who were most severely affected by disaster. My term for this willingness to negotiate knowledge-making mechanisms and the nonnegotiables of disaster reconstruction is *epistemic flexibility*.

4. Acknowledge that scarcity in disaster reconstruction is politically and epistemically produced but not absolute.

In post-Mitch Honduras and post-Katrina New Orleans, there were key instances when expert planners and reconstruction project managers denied disaster survivors the kinds of socio-material arrangements they deemed necessary to produce an affective experience of recovery because the expense of such arrangements would have exceeded institutional cost-benefit analyses. Ironically, aid agencies and institutions (USAID, the Louisiana Recovery Authority) in both post-disaster contexts found spending their aid budgets on time to be one of the most pressing challenges of recovery, and they sometimes failed to do so. In these key instances, ethnographic research revealed an important contradiction: while disaster survivors were routinely denied the resources they needed to reconstruct in a meaningful and sustainable way, institutional actors perceived the abundance of recovery resources as a major challenge of reconstruction efforts (i.e., spending the budget on time). In both cases, power operated on the disaster-affected populations by letting this contradiction go unchallenged and unaddressed. Reconstruction program managers must avoid using cost-benefit analysis as a justification for providing disaster survivors with minimal aid packages (which often amount to a wasteful use of resources as disaster survivors abandon inadequate assistance). Instead, they should work creatively and flexibly to channel resources to those who need it most.

5. Distinguish the body politic in disaster reconstruction—that is, who is the client, who is served.

The ways class, race, gender, and ethnic difference are made, sustained, and affectively experienced in a disaster-affected site play important roles in shaping how a catastrophe's socio-material impacts are distributed and the ease (or difficulty) with which people manage to gain a sense of recovery after a catastrophe. Not only are those people who are most affected by disasters and prolonged reconstruction processes often viewed by local elites with disdain, suspicion, condescension, or judgment, but frequently their ways of speaking, thinking, and space and personhood making evoke averse or skeptical affective reactions in recovery experts, public officials, and commentators.

In post-Mitch Honduras, Samaritan's Purse project managers and local government officials viewed *clase obrera* residents of heavily flooded neighborhoods not as assertive equal partners in the recovery endeavor but as marginal urban populations who, in the moral economy of post-disaster reconstruction, should be grateful recipients of minimal aid packages. In post-Katrina New Orleans, expert planners perceived African American inner-city residents of modest means as people who lacked expertise in urban development and who could only be instructed in neoliberal ways of making and experiencing the city's spaces instead of as equal authors of the city's reconstruction directive—even though the UNOP and the Lambert Plan were legitimized using just such a representation of the planning process. Expert actors and local government officials saw the people most affected by disasters as obstacles to development endeavors rather than as the clients of disaster recovery processes. On the contrary, in New Orleans, the city's Housing Authority and the Department of Housing and Urban Development became the primary clients of expert planners, while in Honduras, the executive leadership of Samaritan's Purse became the main beneficiaries of housing reconstruction programs.

In southern Illinois, because of their suspected political leanings and aesthetic preferences, the rural working-class residents of Olive Branch were themselves seen with suspicion by the very academics and architects who set out to assist them. Disaster recovery experts and institutional actors, therefore, must pay close attention to how their views on class, race, and ethnicity manifest in their own affective preferences as they relate to the various actors and clients involved in disaster mitigation and reconstruction. As Kathleen Stewart (2007) has observed, affect is what happens in the half second when a police officer decides to shoot an African American man for pulling out his wallet or when we fall in love at first sight as someone walks into a room. Disaster reconstruction is not immune from the affective manifestation of habitually inculcated classist, racist, and ethnocentric cultural preferences. Finally, aid workers must reflect on the ethical conflicts of interest that play out when the careers and interests of those who manage and distribute aid start to take precedence over their work in the recovery of disaster-affected communities.

Affect and Disasters

Just as an affective focus helps shed light on the embodied and ecological dimensions of disaster recovery, disaster reconstruction also presents us with a rich ethnographic context for theorizing affect. As I noted in chapter 1, cultural studies and philosophical approaches to affect (exemplified in the work of Brian Massumi) tend to accept the knowledge of neurological science but ignore the contingency and historicity of the sensing body, leading to the universalizing of some bodies and sensory experiences while ignoring others. Emily Martin (2013) has critiqued such tendencies and called for a renewed focus on subjectivity in affect studies.

The various case studies I have presented, however, pose a double challenge to both Massumi and Martin and build a case for an approach to affect that defies the dichotomy between the social construction of subjectivity and the objectivism of neurological science. That dichotomy can only be bridged through an ecological approach where ecology is understood as an ever-shifting and emergent web of co-constitutive meaning-laden relationships between people, their sensing bodies, their meaning-investing capacities, and their surrounding social, built, and natural environments. This book's multisited approach reveals the multiplicity and contingency of affective experience, demonstrating that affect cannot be reduced to a mechanical or biologically determined phenomenon or a social construct. Affect is neither subject nor object but a relational ontology that brings meaning, memory, body, and materiality together over the course of experience.

References

Abu-Lughod, Lila, and Catherine Lutz. 1990. "Introduction: Emotion, Discourse, and the Politics of Everyday Life." In *Language and the Politics of Emotion*, edited by Catherine Lutz and Lisa Abu-Lughod, 1–23. Cambridge: Cambridge University Press.

Adams, Jane, ed. 2002. *Fighting for the Farm: Rural America Transformed.* Philadelphia: University of Pennsylvania Press.

Adams, Richard N. 1970. *Crucifixion by Power: Essays on Guatemalan National Social Structure, 1944–1966.* Austin: University of Texas Press.

Adams, Vincanne. 1998. *Doctors for Democracy: Health Professionals in the Nepal Revolution.* Cambridge: Cambridge University Press.

———. 2013. *Markets of Sorrow, Labors of Faith: New Orleans in the Wake of Katrina.* Durham: Duke University Press.

Agamben, Giorgio. 1998. *Homo Sacer: Sovereign Power and Bare Life.* Stanford: Stanford University Press.

Allen, Greg. 2015. "Ghosts of Katrina Still Haunt New Orleans' Shattered Lower Ninth Ward." *Morning Edition.* National Public Radio, August 18. Accessed December 12, 2015. http://www.npr.org/2015/08/03/42784 4717/ghosts-of-katrina-still-haunt-new-orleans-shattered-lower -ninth-ward.

Allison, Michael E. 2006. "The Transition from Armed Opposition to Electoral Opposition in Central America." *Latin American Politics and Society* 48 (4): 137–62.

AMDC (Alcaldía Municipal del Distrito Central). 1999. *Los Daños a la Capital en Cifras* [Damage to the capital city in numbers]. Tegucigalpa: Alcaldía Municipal del Distrito Central.

Anderson, Mary B. 1994. "Understanding the Disaster-Development Continuum: Gender Analysis Is the Essential Tool." *Focus on Gender* 2:7–10.

Appadurai, Arjun. 1996. *Modernity at Large: Cultural Dimensions of Globalization.* Minneapolis: University of Minnesota Press.

Arce, Alberto, and Norman Long, eds. 2000. *Anthropology, Development, and Modernities: Exploring Discourses, Counter-Tendencies, and Violence.* New York: Routledge.

Audefroy, Joel B., and Bertha N. Cabrera. 2014. "Populations Déplacées par les Désastres et par le Changement Climatique au Mexique" [Populations displaced by disasters and climate change in Mexico]. In *Terres (Dés) humanisées: Ressources et Climat*, edited by Charlotte Bréda, Mélanie Chaplier, Julie Hermesse, and Emanuelle Piccoli, 239–59. Louvain-la-Neuve: Academia-L'Harmattan Press.

Bankoff, Greg, and Dorothea Hilhorst. 2004. "Introduction: Mapping Vunerability." In *Mapping Vulnerability: Disasters, Development, and People*, edited by Greg Bankoff, Georg Frerks, and Dorothea Hilhorst, 1–9. London: Earthscan Press.

Barrios, Roberto E. 2010. "You Found Us Doing This, This Is Our Way: Criminalizing Second Lines, Super Sunday, and Habitus in Post-Katrina New Orleans." *Identities: Global Studies in Culture and Power* 17 (6): 586–612.

———. 2011. "'If You Did Not Grow Up Here, You Cannot Appreciate Living Here': Neoliberalism, Space-time, and Affect in Post-Katrina Recovery Planning." *Human Organization* 70 (2): 118–27.

Barrios, Roberto E., James P. Stansbury, Rosa Palencia, and Marco T. Medina. 2000. "Nutritional Status of Children under 5 Years of Age in Three Hurricane-Affected Areas of Honduras." *Pan American Journal of Public Health* 8 (6): 380–84.

Basch, Linda, Nina Glick Schiller, and Cristina Szanton Blanc. 1994. *Nations Unbound: Transnational Projects, Postcolonial Predicaments, and Deterritorialized Nation-States*. Philadelphia: Gordon and Breach Science.

Baudrillard, Jean. 1995. *Simulacra and Simulation*. Translated by Sheila Faria Glaser. Ann Arbor: University of Michigan Press.

Bauman, Richard, and Charles Briggs. 2003. *Voices of Modernity: Language Ideologies and the Politics of Inequality*. Cambridge: Cambridge University Press.

Beriss, David. 2012. "Red Beans and Rebuilding: An Iconic Dish, Memory and Culture in New Orleans." In *Rice and Beans: A Unique Dish in a Hundred Places*, edited by Richard Wilk and Livia Barbosa, 241–63. London: Berg.

Bhabha, Homi. 2004. *The Location of Culture*. New York: Routledge.

Biersack, Aletta. 1999. "Introduction: From the 'New Ecology' to New Ecologies." *American Anthropologist* 101 (1): 5–18.

Boellstorff, Tom. 2005. *The Gay Archipelago: Sexuality and Nation in Indonesia*. Princeton: Princeton University Press.

———. 2015. "Emergent Coherences." *Cultural Anthropology*, July 21. Accessed June 29, 2016. https://culanth.org/fieldsights/706-emergent-coherences.

Bordo, Susan. 1993. "Feminism, Foucault, and the Politics of the Body." In *Up *against Foucault: Explorations of Some Tensions between Foucault and Feminism*, edited by Caroline Ramazanoglu, 179–203. New York: Routledge.

Bourdieu, Pierre. 1977. *Outline of a Theory of Practice*. Cambridge: Cambridge University Press.

Bradshaw, Sarah. 2001. "Reconstructing Roles and Relations: Women's Participation in Reconstruction in Post-Mitch Nicaragua." *Gender and Development* 9:79–87.

Breunlin, Rachel, and Helen Regis. 2006. "Putting the Ninth Ward on the Map: Race, Place, and Transformation in Desire, New Orleans." *American Anthropologist* 108 (4): 744–64.

Breunlin, Rachel, and Ronald W. Lewis. 2009. *The House of Dance and Feathers: A Museum by Ronald W. Lewis*. New Orleans: University of New Orleans Press.

Briggs, Charles. 2004. "Theorizing Modernity Conspiratorially: Science, Scale, and the Political Economy of Public Discourse in Examinations of a Cholera Epidemic." *American Ethnologist* 3 (2): 164–87.

Briones, Fernando. 2010. "Inundados, Reubicados, y Olvidados: Traslado del Riesgo de Desastres en Motozintla, Chiapas" [Flooded, resettled, and forgotten: Shifting of risk in Motozintla, Chiapas]. *Revista de Ingeniería* 30:132–44.

Brown, Ronald H., D. James Baker, and Elbert W. Friday. 1994. *Natural Disaster Survey Report: The Great Flood of 1993*. Washington DC: U.S. Department of Commerce.

Browne, Katherine E. 2013. "'Culture Brokers' Have Role to Play in Flood Disaster Recovery." *Coloradoan*, October 4. Accessed November 17, 2014. http://www.coloradoan.com/article/20131004/OPINION04/310040 028/Soapbox-Culture-brokers-role-play-flood-disaster recovery. (Link now inactive.)

———. 2015. *Standing in the Need: Culture, Comfort, and Coming Home after Katrina*. Austin: University of Texas Press.

Burdeau, Cain. 2011. "Katrina Homeowners Will Share $62M in HUD Settlement." *Houston Chronicle*, July 6. Accessed November 15, 2014. http://www.chron.com/news/nation-world/article/Katrina-homeowners-will -share-62M-in-HUD-2077732.php.

Burnett, John. 2006. "Larger-Than-Life Sheriff Rules Louisiana Parish." *All Things Considered*. National Public Radio, November 28. Accessed June 10, 2014. http://www.npr.org/templates/story/story.php?storyId=6549329.

Burns, Allan F. 1993. *Maya in Exile: Guatemalans in Florida*. Philadelphia: Temple University Press.

Button, Gregory. 2010. *Disaster Culture: Knowledge and Uncertainty in the Wake of Human and Environmental Catastrophe*. Walnut Creek: Left Coast Press.

Button, Gregory, and Anthony Oliver-Smith. 2008. "Disaster, Displacement, and Employment: Distortion of Labor Markets during Post-Katrina Reconstruction." In *Capitalizing on Catastrophe: Neoliberal Strategies of Disaster Reconstruction*, edited by Nandini Gunewardena and Mark Schuller, 123–46. Lanham MD: Altamira Press.

Caldeira, Teresa, and James Holston. 2005. "State and Urban Space in Brazil: From Modernist Planning to Democratic Interventions." In *Global Assemblages: Technology, Politics, and Ethics as Anthropological Problems*, edited by Aihwa Ong and Stephen J. Collier, 393–416. Malden MA: Wiley-Blackwell.

Camillo, Charles A. 2012. *Divine Providence: The 2011 Flood in the Mississippi River and Tributaries Project*. Vicksburg: Mississippi River Commission.

Campanella, Richard. 2006. *Geographies of New Orleans: Urban Fabrics before the Storm*. Lafayette: Center for Louisiana Studies Press.

Cañizares-Esguerra, Jorge. 2002. *How to Write the History of the New World: Histories, Epistemologies, and Identities in the Eighteenth-Century Atlantic World*. Stanford: Stanford University Press.

Carmack, Robert M., Janine L. Gasco, and Gary H. Gossen. 2006. *The Legacy of Mesoamerica: History and Culture of a Native American Civilization*. 2nd ed. New York: Routledge.

Carrasquillo, Adrian. 2013. "Unlike the Irish, Latinos Can't Assimilate, Says Heritage Immigration Study Co-Author." MSNBC, May 9. Accessed October 3, 2014. http://www.msnbc.com/msnbc/unlike-the-irish-latinos-cant-assimilate-s.

Casa Presidencial de Honduras. 2003. "Operación Contra Maras" [Operation against Maras]. Accessed August 25, 2008. http://www.presidencia.gob.hn/frmControl.aspx?cId=8.

Castañeda, Claudia. 2002. *Figurations: Child, Bodies, Worlds*. Durham: Duke University Press.

CEPAL (Comisión Económica para América Latina y el Caribe). 2008. "Tabasco: Características e Impacto Socioeconómico de las Inundaciones Provocadas a Finales de Octubre y a Comienzos de Noviembre de 2007 por el frente frío número 4." June 16. Accessed October 29, 2014. http://repositorio.cepal.org/handle/11362/25881.

Cernea, Michael M. 1996. "The Risks and Reconstruction Model for Resettling Displaced Populations." *World Development* 25 (10): 1569–87.

Chakrabarty, Dipesh. 2000. *Provincializing Europe: Postcolonial Thought and Historical Difference*. Princeton: Princeton University Press.

Coe, Michael. 2012. *Breaking the Maya Code*. 3rd ed. London: Thames and Hudson.

Collier, Stephen J., and Andrew Lakoff. 2005. "On Regimes of Living." In *Global Assemblages: Technology, Politics, and Ethics as Anthropological Problems*, edited by Aihwa Ong and Stephen J. Collier, 22–39. Malden MA: Wiley-Blackwell.

———. 2015. "Vital Systems Security: Reflexive Biopolitics and the Government of Emergency." *Theory, Culture & Society* 32 (2): 19–51.

Córdoba, Edgar Damián. 2012. "Desastre y Reubicación en Nuevo Juan de Grijalva: Primera Ciudad Autosustentalbe del Mundo." BA thesis, Benemérita Universidad Autónoma de Puebla Publicador.

Cuny, Frederick C. 1983. *Disasters and Development*. New York: Oxford University Press.

CUPA (Center for Urban and Public Affairs). 1995. "Enhancing the Sense of Place in Tremé: Mechanisms for Preserving a Unique, Historic Neighborhood." New Orleans: University of New Orleans Center for Urban and Public Affairs.

Cupples, Julie. 2007. "Gender and Hurricane Mitch: Reconstructing Subjectivities after Disaster." *Disasters* 31:155–75.

D'Ans, André-Marcel. 1998. *Honduras: Emergencia Difícil de una Nación, de un Estado* [Honduras: Difficult emergence of a nation, of a state]. Tegucigalpa: Renal Video Producción.

Davis, Edwin Adams. 1960. *The Story of Louisiana*. Vol. 2. New Orleans: J. F. Hyer.

Dawdy, Shannon Lee. 2008. *Building the Devil's Empire: French Colonial New Orleans*. Chicago: University of Chicago Press.

De Cunzo, Lu Ann. 2001. "On Reforming the 'Fallen' and Beyond: Transforming Continuity at the Magdalen Society of Philadelphia, 1845–1916." *International Journal of Historical Archaeology* 5 (1): 19–43.

Deleuze, Gilles, and Félix Guatarri. 2009. *Anti-Oedipus: Capitalism and Schizophrenia*. New York: Penguin Classics.

DeMello, Margo. 2000. *Bodies of Inscription: A Cultural History of the Modern Tattoo Community*. Durham: Duke University Press.

Diario el Heraldo. 2000a. "Un Cadaver Descuartizado en las Alcantarillas de Centro Penal" [A corpse dismembered in the sewers of penal center]. October 11, sección costa norte, 1.

———. 2000b. "Preparan Decreto para regular Uso de los Vehiculos con Paila" [Decree prepared to regulate use of vehicles with beds]. October 14.

Diario la Tribuna. 2000. "Testigos Fueron Claves: Dictan Prision a Mareros que Descuartizaron un Reo." October 20.

di Leonardo, Micaela. 2008. "Introduction: New Global and American Landscapes of Inequality." In *New Landscapes of Inequality: Neoliberalism and the Erosion of Democracy in America*, edited by Jane L. Collins, Micaela di Leonardo, and Brett Williams, 3–20. Santa Fe: School for Advanced Research Press.

Doss, Erika. 2010. *Memorial Mania: Public Feeling in America*. Chicago: University of Chicago Press.

Eggler, Bruce. 2009. "Nagin Dissolves Hurricane Recovery Office Blakeley Headed." *Times-Picayune*, September 1. Accessed June 27, 2016. http://www.nola.com/politics/index.ssf/2009/09/nagin_dissolves_hurricane_reco.html.

Ehrenreich, Jeffrey David. 2004. "Bodies, Beads, Bones and Feathers: The Masking Tradition of Mardi Gras Indians in New Orleans—a Photo Essay." *City & Society* 16 (1): 117–50.

Enarson, Elaine. 1998. *Surviving Domestic Violence and Disasters*. Vancouver: Freda Center for Research on Violence against Women and Children.

———. 2000. "Gender and Natural Disasters." Working Paper 1. Geneva: InFocus Programme on Crisis Response and Reconstruction, September. Accessed January 1, 2014. http://natlex.ilo.ch/wcmsp5/groups/public/@ed_emp/@emp_ent/@ifp_crisis/documents/publication/wcms_116391.pdf.

Ensor, Marisa. 2009. "Gender Matters in Post-Disaster Reconstruction." In *The Legacy of Hurricane Mitch*, edited by Marisa O. Ensor, 129–55. Tucson: University of Arizona Press.

Escobar, Arturo. 1995. *Encountering Development: The Making and Unmaking of the Third World*. Princeton: Princeton University Press.

———. 1997. "Anthropology and Development." *International Social Science Journal* 49 (4): 497–513.

Euraque, Dario. 1996. *Reinterpreting the Banana Republic: Region and State in Honduras, 1870–1972*. Chapel Hill: University of North Carolina Press.

Fabian, Johannes. 1983. *Time and the Other: How Anthropology Makes Its Object*. New York: Columbia University Press.

Farquhar, Judith. 1994. "Eating Chinese Medicine." *Cultural Anthropology* 9 (4): 471–97.

———. 2002. *Appetites: Food and Sex in Post-Socialist China*. Durham: Duke University Press.

Fassin, Didier. 2013. "On Resentment and *Ressentiment*: The Politics and Ethics of Moral Emotions." *Current Anthropology* 54 (3): 249–67.

Ferguson, James. 1994. *The Anti-Politics Machine: "Development," Depoliticization, and Bureaucratic Power in Lesotho*. Minneapolis: University of Minnesota Press.

———. 1999. *Expectations of Modernity: Myths and Meanings of Urban Life in the Zambian Copperbelt*. Berkeley: University of California Press.

Ferguson, James, and Larry Lohman. 1994. "The Anti-Politics Machine: 'Development' and Bureaucratic Power in Lesotho." *The Ecologist* 24 (5): 176–81.

Fletcher, Laurel E., Phuong Pham, Eric Stover, and Patrick Vinck. 2006. "Rebuilding after Katrina: A Population-Based Study of Labor and Human Rights in New Orleans." Berkeley: Human Rights Center, University of California, June. Accessed December 2012. https://www.researchgate.net/publication/228275531_Rebuilding_after_Katrina_A_Population-Based_Study_of_Labor_and_Human_Rights_in_New_Orleans.

Fortun, Kim. 2001. *Advocacy after Bhopal: Environmentalism, Disaster, New Global Orders*. Chicago: University of Chicago Press.

Foster-Fishman, P., S. J. Pierce, and L. A. Van Egeren. 2009. "Who Participates and Why: Building a Process Model of Citizen Participation." *Health Education & Behavior* 36 (3): 550–69.

Foucault, Michel. 1970. *The Order of Things: An Archaeology of the Human Sciences*. New York: Random House.

———. 1978. *History of Sexuality*. Vol. 1. New York: Random House.

———. 1980. *Power/Knowledge: Selected Interviews and Other Writings, 1972–1977*. Edited by Colin Gordon. New York: Pantheon Books.

———. 1991. "Governmentality." In *The Foucault Effect: Studies in Governmentality*, edited by Graham Burchell, Colin Gordon, and Peter Miller, 87–104. Chicago: University of Chicago Press.

———. 2004. *Security, Territory, Population: Lectures at the College de France, 1977–1978*. Edited by Michel Senellart. New York: Picador.

Fussell, Elizabeth. 2007. "Latino/a Immigrants in Post-Katrina New Orleans: A Research Report." *World on the Move* 13 (2): 2–4.

Gamburd, Michele Ruth. 2013. *The Golden Wave: Culture and Politics after Sri Lanka's Tsunami Disaster*. Bloomington: Indiana University Press.

Ganapati, Nazife E., and Sukumar Ganapati. 2009. "Enabling Participatory Planning in Post-Disaster Contexts: A Case Study of World Bank's Housing Reconstruction in Turkey." *Journal of the American Planning Association* 75 (1): 41–59.

Gaonkar, Dilip, ed. 2001. *Alternative Modernities*. Durham: Duke University Press.

Gibbs, Anna. 2010. "After Affect: Sympathy, Synchrony, and Mimetic Communication." In *The Affect Theory Reader*, edited by Melissa Gregg and Gregory J. Seigworth, 186–205. Durham: Duke University Press.

GNOCDC (Greater New Orleans Community Data Center). 2007. "Pre-Katrina Data Center Web Site." http://www.gnocdc.org/prekatrinasite.html.

Gobierno del Estado de Chiapas. 2013. *Manual de Organización: Instituto de*

Población y Ciudades Rurales. Tuxtla Gutiérrez: Gobierno del Estado de Chiapas Publicador.

Gordillo, Gastón. 2014. *Rubble: The Afterlife of Destruction*. Durham: Duke University Press.

Gray, Peter, and Kendrick Oliver. 2004. *The Memory of Catastrophe*. Manchester: Manchester University Press.

Greet, Pamela. 1994. "Making Good Policy into Good Practice." *Focus on Gender* 2:11–13.

Grimm, Andy. 2014. "Ray Nagin in His Own Words: The Soul of New Orleans, Chocolate City, and the Biggest G——n Crisis in the History of This Country." *Times-Picayune*, September 8. Accessed October 5, 2014. http://www.nola.com/crime/index.ssf/2014/09/ray_nagin_in_his_own_words_the.html.

Guggenheim, Scott, and Michael M. Cernea. 1993. "Anthropological Approaches to Involuntary Resettlement: Policy, Practice, and Theory." In *Anthropological Approaches to Resettlement*, edited by Michael Cernea and Scott Guggenheim, 1–12. Boulder: Westview Press.

Guha, Ranajit. 1983. *Elementary Aspects of Peasant Insurgency in Colonial India*. Durham: Duke University Press.

Gunewardena, Nandini, and Mark Schuller, eds. 2008. *Capitalizing on Catastrophe: Neoliberal Strategies in Disaster Reconstruction*. Lanham: AltaMira Press.

Gupta, Akhil. 2012. *Red Tape Bureaucracy: Bureaucracy, Structural Violence, and Poverty in India*. Durham: Duke University Press.

Gupta, Akhil, and James Ferguson. 1992. "'Beyond Culture': Space, Identity, and the Politics of Difference." *Cultural Anthropology* 7 (1): 6–23.

Gutmann, Matthew. 1996. *The Meanings of Macho: Being a Man in Mexico City*. Berkeley: University of California Press.

Haraway, Donna J. 1997. *Modest_Witness@Second_Millennium.FemaleMan©_Meets_OncoMouse™: Feminism and Technoscience*. New York: Routledge.

Hastrup, Frida. 2011. *Weathering the World: Recovery in the Wake of the Tsunami in a Tamil Fishing Village*. Oxford: Berghahn Books.

Hill, Jonathan D., and Thomas M. Wilson. 2003. "Identity Politics and the Politics of Identities." *Identities: Global Studies in Culture and Power* 10 (1): 1–8.

Hirsch, Arnold, and Joseph Logsdon, eds. 1992. *Creole New Orleans: Race and Americanization*. Baton Rouge: Louisiana State University Press.

Hoffman, Susanna. 1999. "The Regenesis of Traditional Gender Patterns in the Wake of Disaster." In *The Angry Earth: Disaster in Anthropological Perspective*, edited by Anthony Oliver-Smith and Susanna Hoffman, 123–92. New York: Routledge.

Hoffman, Susanna, and Anthony Oliver-Smith. 1999. "Anthropology and the Angry Earth: An Overview." In *The Angry Earth: Disaster in Anthropological Perspective*, edited by Anthony Oliver-Smith and Susanna Hoffman, 1–16. New York: Routledge.

Holston, James. 1989. *The Modernist City: An Anthropological Critique of Brasilia*. Chicago: University of Chicago Press.

Hsu, Frances. 1977. "Role, Affect, and Anthropology." *American Anthropologist* 79:805–8.

Ingold, Tim. 2000. *The Perception of the Environment: Essays on Livelihood, Dwelling, and Skill*. New York: Routledge.

Irazábal, Clara, and Jason Neville. 2007. "Neighbourhoods in the Lead: Grassroots Planning for Social Transformation in Post-Katrina New Orleans?" *Planning, Practice & Research* 22 (2): 131–53.

Isbister, John. 1991. *Promises Not Kept: The Betrayal of Social Change in the Third World*. West Hartford: Kumarian Press.

Jackson, Antoinette. 2011. "Diversifying the Dialogue Post-Katrina—Race, Place, and Displacement in New Orleans, U.S.A." *Transforming Anthropology* 19 (1): 3–16.

Jackson, Deborah Davis. 2011. "Scents of Place: The Dysplacement of a First Nations Community in Canada." *American Anthropologist* 113 (4): 606–18.

Jameson, Fredric. 1991. *Postmodernism, or, the Cultural Logic of Late Capitalism*. Durham: Duke University Press.

Jansen, Kees. 1998. *Political Ecology, Mountain Agriculture, and Knowledge in Honduras*. Amsterdam: Thela.

Johnson, Cedric, ed. 2011. *The Neoliberal Deluge: Hurricane Katrina, Late Capitalism, and the Remaking of New Orleans*. Minneapolis: University of Minnesota Press.

Johnson, Jerah. 1992. "Colonial New Orleans: A Fragment of the Eighteenth-Century French Ethos." In *Creole New Orleans: Race and Americanization*, edited by Arnold R. Hirsch and Joseph Logsdon, 12–57. Baton Rouge: Louisiana State University Press.

Kant, Immanuel. 1996 [1797]. *The Metaphysics of Morals*. Translated and edited by Mary Gregor. Cambridge: Cambridge University Press.

Keane, Webb. 2007. *Christian Moderns: Freedom and Fetish in the Mission Encounter*. Berkeley: University of California Press.

Klein, Naomi. 2007. *The Shock Doctrine: The Rise of Disaster Capitalism*. New York: Picador Press.

Krupa, Michelle. 2007a. "Lunch Trucks Cart Their Food, Tax Dollars to New Orleans." *Times-Picayune*, June 28. http://blog.nola.com/updates/2007/06/lunch_trucks_cart_their_food_t.html.

———. 2007b. "N.O. Plans Leaving Drawing Board." *Times-Picayune*, January 7. Accessed October 2010. http://www.nola.com/news/t-p/metro/index .ssf?/base/news-19/1168156457152530.xml&.xml&cll=1&thispage=1. (Link now inactive.)

Kurwicki, Holden. 2015. "FEMA Buyouts Turn Olive Branch, IL., Neighborhoods into Ghost Towns." WPSD Local 6, April 20. Accessed July 6, 2016. http://www.wpsdlocal6.com/story/28851923/fema-buyouts-turn-olive -branch-il-neighborhoods-into-ghost-towns.

Kusel, Jonathan. 1996. "Well-being in Forest-Dependent Communities, Part I: A New Approach." In *Sierra Nevada Ecosystem Project: Final Report to Congress*, 2:361–73. Davis: University of California, Centers for Water and Wildland Resources.

Lacho, Kenneth J., and Kenneth Fox. 2001. *An Analysis of an Inner-City Neighborhood: Tremé-Past, Present, and Future*. Conway: Small Business Advancement National Center, University of Central Arkansas.

Langford, Jean M. 1999. "Medical Mimesis: Healing Signs of a Cosmopolitan 'Quack.'" *American Ethnologist* 26 (1): 24–46.

Latour, Bruno. 1993a. *The Pasteurization of France*. Translated by Alan Sheridan and John Law. Cambridge: Harvard University Press.

———. 1993b. *We Have Never Been Modern*. Translated by Catherine Porter. Cambridge: Harvard University Press.

———. 1999. *Pandora's Hope: Essays on the Reality of Science Studies*. Cambridge: Harvard University Press.

Lefebvre, Henri. 1992. *The Production of Space*. Oxford: Wiley-Blackwell.

———. 1996. "The Right to the City." In *Writings on Cities*, translated and edited by Eleonore Kofman and Elizabeth Lebas, 147–59. Oxford: Wiley-Blackwell.

Leonard, Mary Delach. 2013. "After Great Flood of '93, Valmeyer, Ill., Retreated to the Bluffs and Found Its Future." *St. Louis Beacon*, July 30. Accessed August 2014. https://www.stlbeacon.org/#!/content/31998 /valmeyer_part_one_072313.

Leonard, Thomas M. 2012. "Vaccaro Brothers Fruit Company." In *Encyclopedia of US-Latin American Relations*, edited by Thomas M. Leonard, Jurgen Buchenau, Graeme Mount, and Kyle Longley, 930–31. Thousand Oaks: Sage.

Lipsitz, George. 1988. "Mardi Gras Indians: Carnival and Counter-Narrative in Black New Orleans." *Cultural Critique* 10:11–121.

———. 2006. "Learning from New Orleans: The Social Warrant of Hostile Privatism and Competitive Consumer Citizenship." *Cultural Anthropology* 21 (3): 451–68.

Lock, Margaret, and Judith Farquhar, eds. 2007. *Beyond the Body Proper: Reading the Anthropology of Material Life*. Durham: Duke University Press.

Lomnitz, Claudio. 2001. *Deep Mexico, Silent Mexico: An Anthropology of Nationalism*. Minneapolis: University of Minnesota Press.

Low, Setha. 2000. *On the Plaza: The Politics of Public Space and Culture*. Austin: University of Texas Press.

———. 2003. *Behind the Gates: Life, Security, and the Pursuit of Happiness in Fortress America*. New York: Routledge.

———. 2009. "Maintaining Whiteness: The Fear of Others and Niceness." *Transforming Anthropology* 17 (2): 72–92.

———. 2011. "Claiming Space for an Engaged Anthropology: Spatial Inequality and Social Exclusion." *American Anthropologist* 113 (3): 389–407.

Low, Setha, and Denise Lawrence-Zúñiga, eds. 2003. *The Anthropology of Space and Place: Locating Culture*. Malden MA: Wiley-Blackwell.

Lutz, Catherine. 1986. "Emotion, Thought, and Estrangement: Emotion as a Cultural Category." *Cultural Anthropology* 1 (3): 287–307.

———. 1988. *Unnatural Emotions: Everyday Sentiments on a Micronesian Atoll and Their Challenge to Western Theory*. Chicago: Chicago University Press.

Macías, Jesús Manuel, ed. 2009. *Investigación Evaluativa de Reubicaciones Humanas por Desastres en México* [Evaluative research of human resettlements due to disaster in Mexico]. Tlalpan: CIESAS.

Mack, Vicki, and Elaine Ortiz. 2013. "Who Lives in New Orleans and the Metro Area Now? Based on 2012 U.S. Census Bureau Data." New Orleans: Greater New Orleans Community Data Center, September 26.

MacLeod, Murdo. 1973. *Spanish Central America: A Socioeconomic History, 1520–1720*. Berkeley: University of California Press.

Makley, Charlene. 2014. "Spectacular Compassion: 'Natural' Disasters and National Mourning in China's Tibet." *Critical Asian Studies* 46 (3): 371–404.

Maldonado, Julie Koppel, Christine Shearer, Robin Bronen, Kristina Peterson, and Heather Lazrus. 2013. "The Impact of Climate Change on Tribal Communities in the US: Displacement, Relocation, and Human Rights." *Climate Change* 120 (3): 601–14.

Mandel, Ruth. 2008. *Cosmopolitan Anxieties: Turkish Challenges to Citizenship and Belonging in Germany*. Durham: Duke University Press.

Manning, Erin. 2012. *Relationscapes: Movement, Art, Philosophy*. Cambridge: MIT Press.

Marchezini, Victor. 2015. "The Biopolitics of Disaster: Power, Discourses, and Practices." *Human Organization* 74 (4): 362–71.

Marino, Elizabeth. 2015. *Fierce Climate, Sacred Ground*. Fairbanks: University of Alaska Press.

Marino, Elizabeth, and Heather Lazrus. 2015. "Migration or Forced Displacement? The Complex Choices of Climate Change and Disaster Migrants in Shishmaref, Alaska and Nanumea, Tuvalu." *Human Organization* 74 (4): 341–50.

Mariscal, Ángeles. 2009. "Alertan Académicos Sobre Plan de Ciudades Rurales en Chiapas" [Academics issue warning about rural cities plan in Chiapas]. *La Jornada*, August 31. Accessed March 13, 2014. http://www .jornada.unam.mx/2009/08/31/estados/028n1est.

Marszalek, Keith. 2007. "City Announces First 17 Target Recovery Zones." *Times-Picayune*, March 29. http://blog.nola.com/updates/2007/03/city _announces_first_17_target.html.

Martin, Emily. 2013. "The Potentiality of Ethnography and the Limits of Affect Theory." *Current Anthropology* 54 (s7): s149–s158.

Masco, Joseph. 2006. *Nuclear Borderlands: The Manhattan Project in Post–Cold War New Mexico*. Princeton: Princeton University Press.

Massumi, Brian. 2002. "The Autonomy of Affect." In *Parables for the Virtual: Movement, Affect, Sensation*, 23–45. Durham: Duke University Press.

———. 2010. "The Future Birth of the Affective Fact: The Political Ontology of Threat." In *The Affect Theory Reader*, edited by Melissa Gregg and Gregory J. Seigworth, 52–70. Durham: Duke University Press.

Maybin, Eileen. 1994. "Forty Seconds That Shook Their World. The 1993 Earthquake in India: 1. Rebuilding Shattered Lives." *Focus on Gender* 2 (1): 34–36.

Mbembe, Achille. 2001. *On the Postcolony*. Berkeley: University of California Press.

McCarthy, Brendan. 2010. "As National Spotlight Shines on Lower Ninth Ward, 5 Years after Katrina Its Residents Feel Forgotten." *Times-Picayune*, August 29. http://www.nola.com/katrina/index.ssf/2010/08/as_national_spot light_shines_o.html.

McClaurin, Irma. 1996. *Women of Belize: Gender and Change in Central America*. New Brunswick: Rutgers University Press.

Mencias, Tomás Andino. 2002. *Las Maras en Honduras: Investigación Sobre Pandillas y Violencia Juvenil*. Tegucigalpa: Save the Children UK.

Mitchell, Timothy. 2002. *Rule of Experts: Egypt, Techno-Politics, Modernity*. Berkeley: University of California Press.

Mohanty, Chandra, Ann Russo, and Lourdes Torres, eds. 1991. *Third World Women and the Politics of Feminism*. Bloomington: Indiana University Press.

Muir, Jim. 2015. "Muir: Corruption Isn't New in Cairo." *Southern Illinoisan*

(Carbondale), September 5. Accessed July 6, 2016. http://thesouthern
.com/news/opinion/editorial/muir/muir-corruption-isn-t-new-in-cairo
/article_b8e94031-7218-5f45-a51d-dfeccfb4ee92.html.

Navaro-Yashin, Yael. 2012. *The Make-Believe Space: Affective Geography in a
Postwar Polity*. Durham: Duke University Press.

Nelson, Diane. 1999a. *A Finger in the Wound: Body Politics in Quincentennial
Guatemala*. Berkeley: University of California Press.

————. 1999b. "Perpetual Creation and Decomposition: Bodies, Gender, and
Desire in the Assumptions of a Guatemalan Discourse of *Mestisaje*."
Journal of Latin American Anthropology 4 (1): 74–111.

New Orleans Renovation (blog). 2006. "Historic Faubourg Treme Association
Takes Aim at Blight, Crime and Grime." September 17. Accessed June 8,
2010. neworleansrenovation.blogspot.com.

New York Times. 1994. "Weddings: Deborah De Moss, Rene Fonseca." January
2. Accessed December 15, 2013. http://www.nytimes.com/1994/01/02
/style/weddings-deborah-de-moss-rene-fonseca.html.

Norris, Fran H., Susan P. Stevens, Betty Pfefferbaum, Karen F. Wyche, and
Rose L Pfefferbaum. 2008. "Community Resilience as a Metaphor, Theory,
Set of Capacities, and Strategy for Disaster Readiness." *American Journal
of Community Psychology* 41 (1–2): 127–50.

Nossiter, Adam. 2007. "Harry Lee, Outspoken Louisiana Sheriff, Dies at 75."
New York Times, October 2. August 13, 2010. http://www.nytimes.com
/2007/10/02/us/02lee.html.

Oliver-Smith, Anthony. 1986. *The Martyred City: Death and Rebirth in the
Andes*. Albuquerque: University of New Mexico Press.

————. 1999. "What Is a Disaster? Anthropological Perspectives on a
Persistent Question." In *The Angry Earth: Disaster in Anthropological
Perspective*, edited by Anthony Oliver-Smith and Susanna Hoffman, 18–34.
New York: Routledge.

————. 2002. "Theorizing Disasters. Nature, Power and Culture." In *Catastrophe & Culture: The Anthropology of Disaster*, edited by Susanna Hoffman
and Anthony Oliver-Smith Hoffman. Santa Fe: School for American
Research Press.

————, ed. 2009. *Development and Dispossession: The Crisis of Forced Displacement and Resettlement*. Santa Fe: School for Advanced Research Press.

Ong, Aihwa. 2005. "Ecologies of Expertise: Assembling Flows, Managing
Citizenship." In *Global Assemblages: Technology, Politics, and Ethics as
Anthropological Problems*, edited by Aihwa Ong and Stephen J. Collier.
Malden MA: Wiley-Blackwell.

Ong, Aihwa, and Stephen J. Collier, eds. 2005. *Global Assemblages: Technol-*

ogy, Politics, and Ethics as Anthropological Problems. Malden MA: Wiley-Blackwell.

ORDA (Office of Recovery and Development Administration). 2008. "Katrina Recovery Projects." New Orleans: New Orleans Office of Recovery and Development Administration.

Pae, Peter. 2005. "Immigrants Flock to New Orleans as Contractors Fight for Workers." *Los Angeles Times*, October 10. Accessed June 30, 2016. http://articles.latimes.com/2005/oct/10/business/fi-migrants10.

PAHO (Pan American Health Organization). 1998. "Impact of Hurricane Mitch on Central America." *Epidemiology Bulletin* 19 (4): 1–13.

Paolisso, Michael, Sarah Gammage, and Linda Casey. 1999. "Gender and Household-Level Responses to Soil Degradation in Honduras." *Human Organization* 58 (3): 261–73.

Pickering, Andrew. 1995. *The Mangle of Practice: Time, Agency, and Science*. Chicago: University of Chicago Press.

———, ed. 2008. *The Mangle in Practice: Science, Society, and Becoming*. Durham: Duke University Press.

Pineda Cruz, Carlos Mauricio. 2005. "Al-Qaeda's Unlikely Allies in Central America." *Terrorism Monitor* 3 (1): 3–4.

Plato. 2003 [380 BCE]. *The Republic*. Translated by Desmond Lee. London: Penguin Books.

Plyer, Allison, and Vicki Mack. 2015. "Neighborhood Recovery Rates: Growth Continues through 2015 in New Orleans Neighborhoods." The Data Center, July 13. Accessed December 12, 2015. http://www.datacenterre search.org/reports_analysis/neighborhood-recovery-rates-growth -continues-through-2015-in-new-orleans-neighborhoods/.

Povinelli, E. 1995. "Do Rocks Listen? The Cultural Politics of Apprehending Aboriginal Australian Labor." *American Anthropologist* 97 (3): 505–18.

———. 2002. *The Cunning of Recognition: Indigenous Alterities and the Making of Australian Multiculturalism*. Durham: Duke University Press.

———. 2006. *Empire of Love: Toward a Theory of Intimacy, Genealogy, and Carnality*. Durham: Duke University Press.

———. 2010. "Shapes of Freedom: An Interview with Elzabeth A. Povinelli." *Alterites* (1): 88–98.

Proust, Marcel. 2006. *Remembrance of Things Past*. Hertfordshire, UK: Wordsworth Editions.

Quinn, Patrick. 2013. "After Devastating Tornado, Town Is Reborn 'Green.'" *USA Today*, April 15. Accessed October 3, 2014. http://www.usatoday.com /story/news/greenhouse/2013/04/13/greensburg-kansas/2078901/.

Rabinow, Paul. 1995. *French Modern: Norms and Forms of the Social Environment*. Chicago: University of Chicago Press.

———. 2005. "Midst Anthropology's Problems." In *Global Assemblages: Technology, Politics, and Ethics as Anthropological Problems*, edited by Aihwa Ong and Stephen J. Collier, 40–54. Malden MA: Wiley-Blackwell.

Raffestin, Claude. 2007. "Could Foucault Have Revolutionized Geography?" In *Space, Knowledge and Power: Foucault and Geography*, edited by Jeremy W. Crampton and Stuart Elden, 129–40. Burlington VT: Ashgate Publishing Company.

Rappaport, Roy A. 1984. *Pigs for the Ancestors: Ritual in the Ecology of a New Guinea People*. Long Grove IL: Waveland Press.

Reckdahl, Katy. 2007. "Culture, Change Collide in Tréme; Some Residents Balk at Musician's Traditional Sendoff." *Times-Picayune*, February 17.

———. 2008. "Police Break up Tremé Jazz Funeral; Mourners Say NOPD Scattered Second-Line." *Times-Picayune*, May 3.

Regis, Helen. 1999. "Second Lines, Minstresly, and the Contested Landscapes of New Orleans Afro-Creole Festivals." *Cultural Anthropology* 14 (4): 472–504.

Rosenberg, Mark B. 1988. "Narcos and Politicos: The Politics of Drug-Trafficking in Honduras." *Journal of Interamerican Studies and World Affairs* 30 (2–3): 143–65.

Rozario, Kevin. 2007. *Culture of Calamity: Disaster and the Making of Modern America*. Chicago: University of Chicago Press.

Said, Edward. 1979. *Orientalism*. New York: Vintage Books.

Schuller, Mark. 2012. *Killing with Kindness: Haiti, International Aid, and NGOs*. New Brunswick: Rutgers University Press.

Schuller, Mark, and Marilyn M. Thomas Houston. 2006. "Introduction: No Place Like Home, No Time Like the Present." In *Homing Devices: The Poor as Targets of Public Housing Policy and Practice*, edited by Marilyn M. Thomas Houston and Mark Schuller, 1–20. Lanham MD: Lexington Books.

Scudder, Thayer, and Elizabeth Colson. 1982. "From Welfare to Development: A Conceptual Framework for the Analysis of Dislocated People." In *Involuntary Migration and Resettlement: The Problems and Responses of Dislocated People*, edited by Art Hansen and Anthony Oliver-Smith, 267–87. Boulder: Westview Press.

Seigworth, Gregory J., and Melissa Gregg. 2010. "An Inventory of Shimmers." In *The Affect Theory Reader*, edited by Melissa Gregg and Gregory J. Seigworth, 1–28. Durham: Duke University Press.

Shapin, Steven, and Simon Schaffer. 1985. *Leviathan and the Air Pump: Hobbes, Boyle, and the Experimental Life*. Princeton: Princeton University Press.

Sherrieb, Kathleen, Fran H. Norris, and Sandro Galea. 2010. "Measuring Capacities for Community Resilience." *Social Indicators Research* 99 (2): 227–47.

Sieder, Rachel. 1995. "Honduras: The Politics of Exception and Military Reformism, 1972–1978." *Journal of Latin American Studies* 27 (1): 99–127.

Simpson, Edward. 2013. *The Political Biography of an Earthquake: Aftermath and Amnesia in Gujarat, India*. London: Hurst.

Snarr, Neil, and Leonard Brown. 1979. "Permanent Post-Disaster Housing in Honduras: Aspects of Vulnerability to Future Disasters." *Disasters* 3 (3): 287–92.

Sorant, Peter, Robert Whelan, and Alma Young. 1984. "City Profile: New Orleans." *Cities*, May, 314–21.

Spinoza, Benedict de. 1994. *The Ethics and Other Works*. Edited and translated by Edwin Curley. Princeton: Princeton University Press.

Stehlik, Daniela, Geoffrey Lawrence, and Ian Gray. 2000. "Gender and Drought: Experiences of Australian Women in the Drought of the 1990s." *Disasters* 24 (1): 38–53.

Stewart, Kathleen. 2007. *Ordinary Affects*. Durham: Duke University Press.

Stoler, Ann Laura. 1995. *Race and the Education of Desire: Foucault's History of Sexuality and the Colonial Order of Things*. Durham: Duke University Press.

Stonich, Susan. 1993. *"I Am Destroying the Land!" The Political Ecology of Poverty and Environmental Destruction in Honduras*. Boulder: Westview Press.

Sutton, David. 2000. "Whole Foods: Revitalization through Everyday Synesthetic Experience." *Anthropology and Humanism* 25 (2): 120–30.

———. 2001. *Remembrance of Repasts: An Anthropology of Food and Memory*. New York: Bloomsbury Academic Press.

Taussig, Michael. 1991. *Shamanism, Colonialism, and the Wild Man: A Study in Terror and Healing*. Chicago: University of Chicago Press.

———. 1993. *Mimesis and Alterity: A Particular History of the Senses*. New York: Routledge.

Thompson, Ginger. 2004. "Tattooed Warriors: The Next Generation; Shuttling between Nations, Latino Gangs Confound the Law." *New York Times*, September 26.

Thrift, Nigel. 2010. "Understanding the Material Practices of Glamour." In *The Affect Theory Reader*, edited by Melissa Gregg and Gregory J. Seigworth, 289–308. Durham: Duke University Press.

Thurner, Mark. 2003. "After Spanish Rule: Writing Another After." In *After Spanish Rule: Postcolonial Predicaments of the Americas*, edited by Mark Thurner and Andrés Guerrero, 12–57. Durham: Duke University Press.

Tierney, Kathleen, and Anthony Oliver-Smith. 2012. "Social Dimensions of

Disaster Recovery." *International Journal of Mass Emergencies and Disasters* 30 (2): 123–46.

Times-Picayune. 2006. "Editorial: Not Coming Together." June 18. Accessed November 21, 2014. http://www.nola.com/news/t-p/metro/index.ssf ?/base/news-3/1150614069127560. xml&coll=1>. (Link now inactive.)

———. 2011. "1940: Charity Hospital in New Orleans Is Fully Occupied." November 13. Accessed November 20, 2014. http://www.nola.com/175 years/index.ssf/2011/11/1940_charity_hospital_in_new_o.html.

Tobin, Grant, and Linda Whiteford. 2002. "Community Resilience and Volcano Hazard: The Eruption of Tungurahua and Evacuation of the Faldas in Ecuador." *Disasters* 26 (1): 28–48.

Toledano, Roulhac, and Mary Louise Christovich, with Betsy Swanson and Robin Von Breton Derbes. 1980. *New Orleans Architecture*. Vol. 6, *Faubourg Tremé and the Bayou Road: North Rampart Street to North Broad Street, Canal Street to St Bernard Avenue*. Gretna LA: Pelican Publishing.

Tsing, Anna L. 2005. *Friction: An Ethnography of Global Connection*. Princeton: Princeton University Press.

Ullberg, Susann. 2013. *Watermarks: Urban Flooding and Memoryscape in Argentina*. Stockholm: Stockholm University Press.

USACE (United States Army Corps of Engineers). 2014. "MRC History." Accessed December 11, 2014. http://www.mvd.usace.army.mil/About/Mississippi RiverCommission(MRC)/History.aspx.

U.S. Census. 2016. United States Census Bureau Website. Accessed July 6, 2016. http://www.census.gov/search-results.html?q=Greensburg+city %2C+KS&page=1&stateGeo=none&searchtype=web.

Walker, Bridget. 1994. "Editorial: Women and Emergencies." *Focus on Gender* 2:2–6.

Waller, Mark. 2007. "Jefferson Bans Taqueria Trucks." *Times-Picayune*, June 20. Accessed December 2012. http://blog.nola.com/times-picayune/2007 /6/jefferson_bans_taqueria_trucks.html.

Way, J. T. 2012. *The Mayan in the Mall: Globalization, Development, and the Making of Modern Guatemala*. Durham: Duke University Press.

White, Sarah. 1997. "Men, Masculinities, and the Politics of Development." *Gender and Development* 5 (2): 14–22.

Index

Page numbers in italics refer to illustrations.

In the Anthropology of Contemporary North America series:

America's Digital Army: Games at Work and War
Robertson L. Allen

Governing Affect: Neoliberalism and Disaster Reconstruction
Roberto E. Barrios

Holding On: African American Women Surviving HIV/AIDS
Alyson O'Daniel

Rebuilding Shattered Worlds: Creating Community by Voicing the Past
Andrea L. Smith and Anna Eisenstein

To order or obtain more information on these or other University of Nebraska Press titles, visit nebraskapress.unl.edu.